Brigade Commander
Georgii Samoilovich Isserson

THE EVOLUTION
OF
OPERATIONAL ART

Translated by Bruce W. Menning

Combat Studies Institute Press
US Army Combined Arms Center
Fort Leavenworth, Kansas

Foreword

One can argue that the development of true doctrine required the formal adoption of the concept of operational art. Prior to the Great War, no army in the world possessed a codified body of thought that enabled senior military commanders to visualize the aggregate effects of tactical engagements across time and space. By 1918, after a dramatic revision of drill regulations into something approaching true doctrine, the German army was furthest in realizing this goal. Ultimately, though, the Germans could not translate tactical success into strategic victory because they could not resource military operations in sufficient depth to render local successes decisive. Understanding that the character of warfare in 1918 was radically different from 1914 would have enabled Ludendorff to see the flaws in the MICHAEL offensives and perhaps mitigate them. And although the interwar German Army spent a great deal of effort reflecting on the lessons of 1914-1918, German understanding of the operational art remained incomplete.

The separate and unequal Allied efforts against Nazi Germany in World War II, followed immediately by the superpower competition of the Cold War, created a significant gap in American officers' understanding of the factors that contributed to Soviet victories on the Eastern Front. As a result, in the decades following the war the concept of "operational art" was recognized and adopted by the US Army almost as a proprietary creation. In the 1990s, however, Western military historians and theorists discovered that the Soviets had gotten there first.

Bruce Menning's translation of Georgii Samoilovich Isserson's 1936 treatise *The Evolution of Operational Art* is the best example available of the distillation of Soviet military thought before the Second World War. Isserson, Tukhachevsky, Shaposhnikov, and others like them were founding members of a focused military Enlightenment whose goal was to change the way armies and leaders thought about war. Moreover, unlike contemporaries such as B.H. Liddell Hart or Billy Mitchell, they had the opportunity to build their ideas into the modern Soviet Army and see their doctrine survive despite the existential challenges of Stalin's purges and the German invasion. I commend this work to you as a foundational text, one to which I hope you will refer repeatedly throughout your career.

Thomas E. Hanson
Lieutenant Colonel, Infantry
Director, Combat Studies Institute

Contents   *page*

Figures   *page*

Translator's Note

Georgii Samoilovich Isserson's *The Evolution of Operational Art* is a military classic that has long remained inaccessible to non-Russian readers. More than a mere apologia for the Soviet concept of deep battle/operation, this book constitutes a military-intellectual tour de force with its critical analysis of evolving military art in historical-theoretical perspective. The book is also an exercise in military foresight on the nature of future war. In fact, Isserson's final conclusions (page 105-111) could well be understood as a theoretical template for the way that large-scale operations actually unfolded during 1941-45 on the Eastern Front.

More importantly in long-term perspective, Isserson's examination of the prime variables within modern forms for the operation (to use his phraseology) invites the reader to ponder the changing impact and implications of key influences on the evolution of military art. These variables are as modern as today, and they include politico-ideological context, force structures and correlations, command and control, space and depth, time and timing, technology, and technique. On one level, Isserson addresses the challenges inherent in 1930s-vintage future war. On another level, his treatment of overarching issues extends well beyond his time. For example, his examination of Moltke's quandary after Sedan in 1870 (page 56) appears fully appropriate to an analysis of post-2001 US- led operations in Afghanistan and Iraq.

Translation is as much art as science. Russian words and phrases often have no genuine English equivalents, while Russian military terminology is usually quite precise. For example, the Russian word boi usually means "battle," but with distinct tactical connotations. However, the English equivalent can mean anything from a skirmish to the 1914 confrontation on the Marne. In this and related cases, I have tried to apply common sense standards for translation, differentiating, for example, between "battle" (usually small-scale and tactical) and "main battle" (Russian *srazhenie* with large-scale operational and probable strategic implications). In all instances the governing principle has been faithfulness to perceptions of the author's intent.

Brigade Commander Isserson's *The Evolution of Operational Art* treats the main issues of operational art in historical and theoretical perspective. In particular, he critically analyzes the operational heritage of the past, including the age of Napoleon, the epoch of the Franco-Prussian War of 1870-1871, and the era of the imperialist war [the World War]. The author outlines contours for the resolution of operational questions under conditions of future revolutionary-class war, when the deep operation for destruction will become the main form of operation. The work contains some disputable propositions which should stimulate further analysis of the author's ideas. This book was intended primarily for senior officers and high-ranking commanders in the Workers and Peasants Red Army and for students within the senior service academies.

Bruce W. Menning

Introductory Essay

In the 1978 motion picture, *Coming Home,* a U. S. Marine Corps officer, played by Bruce Dern and his wife, played by Jane Fonda, dine together at a Tokyo R & R location. The scene culminates with Fonda asking the inevitable what-was-it-like question. Dern's response is telling: "What was it *like*? It wasn't *like* anything." Or, one can imagine an analogous—but more apocryphal—scene cast with different characters: Ulysses S. Grant and Napoleon Bonaparte relaxing somewhere over tumblers of whiskey and brandy. Napoleon asks Grant, "So, what was the American Civil War like?" Grant pauses, thoughtfully sipping his bourbon, and replies, "*Like*? It wasn't *like* anything." Perhaps only the penetrating lens of military theory can capture the force of Grant's declaration. By Grant's time, the Industrial Revolution had ensured that the art of waging war would be unlike anything during the preceding 2,500 years. By the 1860s, the industrialization of warfare had led not only to the transformation of military art but also in many ways to the very transformation of civilization. Transformation was changing the essential and defining "DNA" of armed conflict. However, the central quality of this transformation has often eluded modern military historians, whose methodology usually focuses on descriptive continuity at the expense of qualitative change.

An important exception to this rule occurred among a small group of Soviet military theorists who wrote during the 1920s and 1930s. They included M. N. Tukhachevsky, B. M. Shaposhnikov, M.V. Frunze, A. A. Svechin, V. K. Triandafillov and G. S. Isserson. Of these—and there were others as well—Isserson was perhaps the most important in arguing and articulating a coherent response to the imaginary dialogue between Grant and Napoleon: If warfare as a consequence of the Industrial Revolution was unlike anything else that preceded it, what did this momentous development mean to the military practitioner, to the operator?

The Collapse of the Classical Paradigm

The destruction of the Tsarist regime in 1917 profoundly affected virtually all traditional Russian institutions, including the military. In many respects the two revolutions of 1917 merely completed the disintegration of the Russian military that had already begun with the start of the Great War. Along with the physical destruction of the old Imperial Army, war and revolution also destroyed much of the former intellectual fabric which had afforded a coherent reference for visualizing the conduct of military operations in modern context. As the Russo-Japanese War of 1904-05 had begun to suggest, and as the Great War was to affirm, the theoretical fabric of Russian military thinking had been spun from fundamentally flawed premises, many of which were irrelevant to the new conditions of industrialized warfare.[1] Many of the same forces that had swept away the old military order in 1917 also fueled the impulse to challenge outmoded and discredited ideas of warfare. Over the two decades after the Bolshevik coup, a handful of Soviet military theorists, including G. S. Isserson, set aside old concepts to derive a new understanding of contemporary war and military art. Isserson's work, *The Evolution of Operational Art,* embodied much of the essence of this movement. Political revolutions often cast aside not only old regimes but also old and seemingly outmoded ideas.[2] In the military realm, the Russian Revolution provided the stimulus for theorists to attack old intellectual shibboleths on the basis of fresh insight and proximate historical experience. With the Great War in the recent past, and even as the Russian Civil War raged, Soviet theorists began to subject traditional military verities to intense scrutiny and criticism. There gradually emerged a new visualization of land warfare appropriate to the changing means and methods of the industrialized twentieth century. By the late 1920s and early 1930s the cumulative work of Soviet theorists added up to a revolution in military thought.

After the Napoleonic Wars, the pace of military revolution quickened dramatically. Advances in technology and altered organization were in the vanguard of change. Initially, the technological transition was most evident in the realm of tactics.[3] The most important technological advance was in the appearance of the rifled musket. Its lethal impact led directly to the expansion of the battlefield. Related innovations included the development of smokeless powder, barbed wire, a reliable breech-loading system, indirect-fire artillery and the machinegun.

By the end of the nineteenth century a field army could control a much broader expanse of frontage than the typical Napoleonic army. Thanks to improvements in artillery indirect fire, the advent of modern means of communication, and the widespread adoption of rifles and machineguns, it became impossible to compress army formations into one compact and dense mass, as had been the case on the Napoleonic battlefield. Altered conditions, as we shall see, led to a revolution in force-on-force deployments that Isserson would label the "strategy of the continuous front." These laterally-extended deployments stood in sharp contrast with the dense, compact deployments on the massed battlefields of Napoleon. Isserson identified these battlefields as the embodiment of the "strategy of a single point."

A century after Napoleon, the Great War witnessed the widespread use of barbed wire and entrenchments. These new realities, together with the appearance of rapid fire artillery and the reappearance of higher troop densities, pressed the "strategy of the continuous front" to its culmination and resulted in a stalemate persisting for four years. Stalemate arose largely from a failure to recognize that the underlying incremental changes wrought by emergent industrial technology and organization added up to a wholly new and fundamentally changed military reality. When the Great War began, most belligerents believed in the possibility of a quick decisive military victory in the Napoleonic mold. Instead, a long war of exhaustion ensued.

It would be easy to dismiss the contending generals and their civilian masters as incompetents or dunderheads, but this charge misses the point. In fact, most Great War generals were often quite effective—by Napoleonic standards. However, by 1914, and even by the American Civil War, Napoleon's form of war—even as expanded and embellished by his heirs—had become irrelevant. In the West theorists were slow to perceive the true nature of the revolution in warfare that had been visited upon them. East of the Elbe and the Vistula, however, new concepts and theories of war were spinning from the minds not of a few clever individuals, but from a small school of officers who had experienced the same struggle by attrition. One of the most stunning and revolutionary of these Soviet theoretical concepts was called *operativnoe iskusstvo,* or *operational art.*

Military Theory and the Operational Turn

Changes in objective reality often require fundamental changes in our subjective thinking if we are to survive successfully in a dynamic and often hostile world. For millennia, our five senses informed us about the environment around us. With the advent of culture and civilization about 5,000 years ago, our senses of taste, touch, hearing, smell and especially vision verged on being overwhelmed by abstract ideas like justice, excellence, freedom and beauty. With abstraction came writing and a further distancing of concept from senses. The Greek philosophers sought to synthesize the intellectual side of human experience and to organize abstract thinking under the rubric of *theoria* as *detached speculation.* This synthesis stood intact until the sixteenth-century, when scientists like Nikolaus Copernicus revolutionized thinking by redefining theory itself. Whereas the Greeks and medieval schoolmen saw theory as detached speculation, practical scientists like Copernicus, Galileo, DaVinci and others viewed theory as a way to *visualize the insensible.* Copernicus, for example, used this new mode of theoretically structured observation to overturn the obsolescent Ptolemaic, or geocentric, cosmology. Using theory as a visualizing tool, Copernicus was able to re-imagine a sun-centered cosmos *without direct reliance* on his five senses. Instead, his theoretical concepts stood as sensible surrogates for the senses, setting the stage for a re-visualized cosmological paradigm.

In much the same way Soviet theorists like Isserson re-visualized the nature of modern warfare that began with the overthrow of the classical Napoleonic paradigm. Under the classical model, which held sway for most of recorded history, war was visualized on two levels or planes of abstraction: the strategic and the tactical. The sudden emergence of the Industrial Revolution ruptured the old Napoleonic cut and prompted a tripartite re-conceptualization. Mostly under Soviet intellectual tutelage a new paradigm called for an intermediate, or *operational level,* and its meaning and implications were articulated in the 1920s and 1930s.

The historical importance of this theoretical contribution cannot be stressed too much. In fundamental terms, the emphasis on an operational level of military actions sought to overcome the tactical orientation that persists in military thinking to this day. This tactical inclination so endures that it constitutes a kind of institutional learning disability. A general officer of the Great War era, for instance, rose professionally from a young subaltern to the height of his profession with a deeply rooted outlook susceptible to a particular expectation bias. In turn, this bias fostered an outlook that viewed the officer's expanding professional horizon as simply a larger version of his previous limited and limiting tactical military experience. Of course military education helped shape this bias with an emphasis on the importance of fighting and winning the decisive Napoleonic battle. Thus, the complex operations and campaigns of the Great War were visualized on maps as Napoleonic battles—but with bigger arrows. Yet upon reflection, these same generals would have denied such preconceptions. For Soviet theorists such denial was a fundamental difficulty: a conceptual influence so subtle and unnamed that it was baffling to express succinctly and perplexing to demonstrate clearly in its pervasive effects. Theorists were left instead with vestiges and intimations of bias presence and influence. The immediate effect, the Soviets realized, was to produce a kind of "tacticization" of operations and campaigns, and even of war itself.

Hints of what we might call the *tactical bias* can be found in at least four areas of cognition. First, there is the way we logically structure the world as we see and imagine it. This perspective is grounded more or less in some version of the past. To some extent additional experience and education help to overcome entrenched expectations from the personal past, but nonetheless a logical fallacy from our early tactical experience becomes part of the very core of our thinking. Logicians might call this the *fallacy of genetic composition:* a false belief that what is true of our tactical past will hold true of our expanded operational and even strategic future. Generals of the Great War applied this fallacy on a routine basis, because the larger world of their later military experience was beyond the direct privileged and privately sensible view of the familiar tactical world witnessed by them as novices and apprentices in the military profession. When overwhelmed by the new and revolutionary diversity of industrialized warfare, the generals tended to paper over their ignorance of the operational and strategic *whole* by defining it in terms of the known and tangible remnants of the tactical *parts*. In contrast, Soviets theorists came to realize that operational art was also *a way of thinking* designed to overcome the fallacy of genetic composition—the tendency to structure cognitively the whole of warfare from the sum of its individual tactical parts.

Second, there was the way in which classical generals were educated. The levels of war remain levels of abstraction, but also uniquely distinct and coherent modes of thought that constitute paradigms in their own right. Other fields, such as theater and film, with a producer-director-actor hierarchy, offer similar planes of abstraction. These levels or planes emerge, often quite suddenly and seemingly arbitrarily, from conceptual necessity. In the military sphere, even today we train almost exclusively to the tactical level of abstraction; it is the easiest to teach and the easiest to learn; it is also the easiest to engineer, short of the rigors of actual practice. The products of such indoctrinated learning systems are, in the words of Walter Klein, "sly and flexible, not so much educated as wised-up."[4] An army of "wised-up" tacticians may make great problem-solvers in an immediate sense, but they also make poor critical thinkers. As generals, they seldom rise above practical application and neglect the cognitive spheres of analysis, synthesis, discernment, appreciation and judgment. The "wised-up" learner is conceptually blind to other, higher levels of abstraction; he cannot see them, because he lacks the theoretical tools of vision and visualization which can only be acquired through serious learning and education. It should come as no surprise then that the Soviet theoretical revolution went hand in hand with a transformation of military education.

Third, there was the question of leadership in military organizations. It became an important challenge for theorists like Isserson. Throughout the classical period, spanning centuries, commanders led using what might be termed the "heroic" model of leadership, creating purpose, direction and motivation through direct physical and sensible presence.[5] Heroic leadership is tactical leadership as old as Achilles and the *Iliad,* a necessary accompaniment to the tactical bias in fighting the decisive climactic battle. The emergence of operational art suggested a new emphasis on intellectual leadership, an autonomous style perhaps first embodied in Odysseus, a more reflective counterpart to Achilles. Under the operational paradigm, which stressed the reality of the massively distributed operation, the leader could no longer impose his will directly on a disaggregated force. Instead, he had to influence the outcome through the force and clarity of his operational vision.

Fourth and finally, there was the perennial institutional assault on military theory that the Soviet thinkers had to overcome. As in most conservative institutions, militaries are typically and unabashedly anti-intellectual. Yet sound military theory was central to the development and elaboration of operational art in the 1920s and 1930s. As Arthur Conan Doyle has Sherlock Holmes say to Dr. Watson: "I have trained myself to notice what I see." Operational theory is about educating the observer to notice what he *visualizes,* so he can make critical judgments about the theater of operations and act—or not act—with deliberation. As such, operational art entails a method of education that strives to overcome the tactical bias in military institutions.

The Soviet endeavor to articulate a new paradigm of warfare was therefore founded primarily upon a series of solid theoretical arguments, rather than through direct experience or through elaboration of a historical narrative. The broader argument was that the evolution of warfare in the West followed a similar process of transformation in human thinking as when, for instance, the revolution in writing changed the very way people thought. During the military transformation, the *tactical bias* remained irrevocably and deeply embedded in the military culture and consciousness. There were, however, certain historical and theoretical glimmerings in which the *tactical bias,* while not fully recognized, was at least challenged. In the classical paradigm, with the Napoleonic Wars as its last full historical expression, warfare had been conceived as a tactical unity for at least two millennia. Tactics, coming from the Greek word which meant an "ordering" or "arrangement" of troops, was the primary level of military abstraction; indeed, it was the only one. By Napoleon's time warfare had been re-conceptualized according to a bifurcated abstraction, with strategy as the *subject* of war and tactics as its *object.* The Industrial Revolution, a proper military revolution in its own right, witnessed the emergence of operational art as the *mediating,* integrative synthesis standing between modern strategy and tactics. Later nineteenth and early twentieth century warfare brought two streams of conceptual development—strategy and tactics—to a higher cognitive plane with a new grand synthesis: *the operational artist emerges as a synthetic mediator whose campaign concept becomes the fundamental referent for the strategic war designer and the tactical battle planner.* These theoretical notions flourished during perhaps the most dynamic period of military thought. The golden age of military thinking in the 1920s and 1930s reached its full culmination and flowering with G. S. Isserson's most prominent work, *The Evolution of Operational Art.*

Beginnings

Georgii Samoilovich Isserson was born on 16 June 1898 in Kaunas, on the Nieman River, and a city today the second largest in Lithuania. In late 1916 he enrolled in a law program at Petrograd University, but was conscripted in early 1917, as the Russian military scrambled to find recruits for its badly mauled armies. The swift collapse of Russia led to an abrupt termination of Isserson's imperial military career. In 1918, he found employment as a private secretary for the Petrograd Printers' Union. After only a month, just four days after his twentieth birthday, he volunteered for active service in the Workers' and Peasants' Red Army (RKKA) of the new Soviet state. Isserson's career in the RKKA shaped his commitment and dedication to Communism. The chaotic events of the Russo-Polish War briefly conspired against him, when in August 1920 he was interned in East Prussia along with 80,000 of his comrades. After his release in November he began seminal military studies in the newly established RKKA General Staff Academy. During 1921, there was a fundamental reorganization of the Academy's curriculum under the intellectual aegis of M. N. Tukhachevsky. He renamed the institution the RKKA Military Academy, extending the term of studies from two to three years and opening up enrollment to line officers in addition to the previously enrolled staff officers. Under Tukhachevsky the curriculum thoroughly assimilated Marxist-Leninist historical theory, especially dialectical materialism, to explain military evolution and transformation. Isserson would rely heavily on this framework to visualize the collapse of the classical military paradigm and the emergence of operational art.

His time at the Academy coincided with the first reformulation of the classical paradigm. Nikolai Efimovich Varfolomeev was one of the foremost military theorists teaching at the Academy. He understood that the old Napoleonic "cut" of strategy and tactics was no longer adequate for proper visualization of modern warfare. Borrowing from A. A. Svechin's broad discussion of strategy, Varfolomeev found it useful by the 1923-1924 academic year to conceptualize warfare abstractly with strategy as the organizing frame for war as a whole, with operational art for integrating disparate tactical actions into a unified operation, and with tactics as the art of the engagement.

Isserson quickly assimilated the operational paradigm of the RKKA academy. Toward the end of his academic tenure he began to develop a serious interest in critical military history. Relying heavily on a facile command of German from his formative years in Kaunas, Isserson turned his attention to an intensive study of German military operations in the Great War. He subjected to practical scrutiny the German failure to comprehend fully the emergence of the operational conditions that had begun to define modern war, first at the battle of Tannenberg, and then in 1918 during the great spring offensive on the Western Front.

During various low-level military postings through the 1920s, Isserson continued his writings and ruminations. By the end of the decade Stalin had seized complete control of the revolutionary Soviet state, moving it sharply to the left by means of aggressive policies for massive industrialization and collectivization. These policies fundamentally changed the entire strategic fabric of the Soviet Union. The transformed nature of the state demanded new ways of conceptualizing and conducting modern operations.

Continuous Front to Deep Battle

In the fall of 1929 Isserson moved from the provinces to a new and important appointment. On 7 October 1929, after having made his mark with military historical writings, he was appointed instructor to the renamed Frunze Military Academy. The academic environment at the Frunze created precisely the proper atmosphere where an intelligent veteran officer could transform himself into a seasoned military theorist. By 1929 the leadership of the Red Army had also began consolidating its military theory into a highly refined reconceptualization of land warfare based on detailed studies of the most recent wars. During the first stage of this development, Soviet theorists, most notably Tukhachevsky and V. K. Triandafillov, formalized the "broad front" concept, which envisioned armies locked in a tactical clinch across an extended deployment front. Recent history demonstrated that contending armies would grapple with one another until one or the other had achieved some sort of breakthrough, allowing for cavalry exploitation or, with the new industrial technology, motorized and mechanized forces. The fielding of new technology led to the further development of a second concept, *deepening,* from which sprang the idea of the *deep battle.*

The deep battle idea was already reflected tactically in the 1929 Field Service Regulations (PU-29). The *deepening* idea was a natural evolution of the *broad front* concept. Soviet thinkers understood that the linear broad front would quickly *thicken* and *deepen* as new technologies were employed, including tanks, aircraft, airborne troops and long-range mobile artillery. This new technology was just under development during Stalin's first massive Five Year Plan (1928-1932). One observer describes the transition as follows:

> Probably the turning point between 'broad front' and 'deep battle' came when the need to reinforce the effort on the main axes led to a deliberate thinning out of the troops in the other sectors until, divisions apart, they came to assume a holding role rather than an offensive one. The conceptual change is perhaps the selection of the main axes *in advance* rather than in response to the course of battle. This did not of course prevent the further reinforcement of a successful thrust at the expense of less successful ones.[6]

Previously, the typical classical battle found dense monolithic armies struggling in a kind of slaughterhouse seldom more than four miles square. This battle, often a decisive battle of annihilation, usually decided the campaign and even the war itself. Under modern industrialized conditions, as already noted, the classical battlefield had begun to spread out *laterally* to create an increasingly *broad front*. By the fall of 1914 the entire Western Front became a linear battlefield—or *broad front*—several hundred miles long. The question Soviet theorists tried to answer was, what made the new configuration *both* a *broad* front and a *deep* battle?

The newly emerging posture of the defender meant that he could deploy his forces into a lateral, broad front in order to defend the whole of his national borders. Securing borders also meant securing the key industrial heartland and therefore the economic sustainment of the nation in the likely event of a protracted war of exhaustion. The defender could also *thicken* his defensive posture by adding several fortified belts and extensive fortifications. Thus, the attacker was confronted with an enemy defense that was both *deep* and *broad*. To achieve decisive penetration, Soviet theorists began to understand that in the offense a breakthrough might occur only with a simultaneous attack through the *depth* of the defense. Otherwise, the attacker would have to "gnaw" his way through the defense at a rate faster than defending foot soldiers could reinforce the point of penetration and breakthrough. But thanks to the development of tanks, aviation, motorized and mechanized infantry, and assault artillery, simultaneous attack through the enemy defense, often 12-15 kilometers deep, could be achieved with an intense cycle of destruction-suppression-paralysis and deep tactical maneuver.

These new theoretical insights were later formalized in the Field Service Regulations of 1936 (PU-36), of which Isserson was editor and coordinating author. He wrote Chapter 6 on the Encounter Battle. Stalin's success with the first two Five Year Plans ensured that the military technology would be largely on hand to resource properly the emerging Deep Battle tactical doctrine.

The Deep Operation

On 26 January 1926, Tukhachevsky promulgated Directive Number 20.030, formally ordaining "A special study concerning the character of future war." The study came as a consequence of the Fourteenth Party Congress, which "formulated a general course for party and Soviet authority on the industrialization of the country" as well as securing the economic and industrial basis for the defense of the Soviet Union.[7] In this lengthy study Tukhachevsky, along with co-authors Ia. K. Berzin, A. N. Nikonov and Ia. M. Zhigur, concluded that, "It is essential to conduct a series of successive operations which are appropriately distributed in space and time. By a combination of a series of operations, it is essential to force the enemy to exhaust its material and human resources or cause the enemy to accept battle by the main mass of troops under disadvantageous conditions and eliminate them."[8]

The deepening idea was taken to the operational level of abstraction in 1929 with the publication of V. K. Triandafillov's *The Nature of the Operations of Modern Armies.* Many today still regard Triandafillov as the father of deep operations.[9] On 12 July 1931 he died tragically in a plane crash. Before his death, Triandafillov wanted to advance his ideas even farther; he had quickly recognized the implications of emerging industrial technology and the increasing mechanization of the Red Army. As something of a testament to the ill-fated theorist, within a few days after the crash a Department of Operations was created at the Frunze Academy. P. I. Vakulich was appointed department head, with Isserson as his deputy. In September 1932, Isserson became department head.

It was during this dynamic and creative period that the 34-year-old Isserson assumed the theoretical mantle of Triandafillov and wrote *The Evolution of Operational Art.* He sought in part to extend the deepening theories developed by Tukhachevsky, and especially Triandafillov, but he also sought his own independent theoretical voice. His initial formulation was presented in February 1931 as an in-house Academy publication under the informal title "The Deep Strategy as the Next Stage in the Evolution of Military Art." It was published the following year in Moscow. (The edition you currently hold in your hands is the revised and expanded second edition published in 1937 by the State Military Publishing House of the USSR People's Defense Commissariat). The first part of the book examines "The Operational Heritage of the Past," a study of the operation and the emergence of operational art. Using an evolutionary framework, Isserson points out that,

> Before the World War, military art admitted only two main elements: strategy as teaching about war, and tactics as teaching about battle. This bifurcated understanding only demonstrated once again how far military theory lagged behind practice.
>
> Even in the second half of the nineteenth century, the evolution of the forms for armed combat exceeded the bounds of this understanding of strategy and tactics. Armed conflict gave birth to a whole chain of combat actions that stretched across a front line and were distributed in depth [discussed earlier in this introduction as broad front and deep battle]. These actions exceeded the limits of battle and these could not be subsumed into tactics. Because these actions did not embrace the phenomenon of war as a whole, they could not be treated as the teaching of strategy on war. [page 13]

Battle itself lost its classical meaning, and,

> in theory, there opened a considerable gap between strategy and tactics, and this gap in the practice of armed combat was filled by real phenomena of great scope and content. These phenomena required a new understanding which emerged only after the World War under the rubric of operational art as instruction on operations.[page 13][10]

The Napoleonic "strategy of a single point" gave way to an age of "linear strategy:"

> A series of new phenomena [which] entered into the unfolding of military events within a theater of war, and these phenomena both exceeded the limits of a single battlefield as a point and outgrew the framework of tactics. Once armies began to enter combat across a broad line, combat efforts became distributed across

a front, and the battle was no longer linked to a single point, but to various points scattered along the front. The main feature of armed conflict during the second half of the nineteenth century was the fact that the single point of Napoleon's era broke down into a series of separate points dispersed in space. [page 19]

The continuous front was fed from the interior of the contending states by their massed industrialized resources and infrastructure. Economic wealth was mobilized at the strategic level and projected along the new railroad infrastructure, which had rapidly developed during the nineteenth century.

By the beginning of the twentieth century, Isserson argued, "An operation *took the shape of a chain of combat efforts along a continuous front, linked in depth, and united by the general intent of defeating or resisting the enemy.*" [page 26][11] Out of this environment arose operational art as a new mediating level of military abstraction synthesizing and linking the strategic mobilization of industrial materiel with that of the tactics of combat action. In other words,

The challenge [of operational art] was to make the chain of combat efforts a highly efficient system coordinated purposefully and sequentially along the front and throughout the depths to bring about enemy's [strategic] defeat. For operational art, the solution for this problem involved contending with the new and complex problem of controlling armies deployed as a continuous front along a single line. [page 26]

But as Isserson goes on to demonstrate, armies failed to develop adequate means of command and control to carry operational art to its full creative potential, thus rendering insoluble the problem of conducting an operational breakthrough against an enemy's front.

The second part of Isserson's book, "The Foundations of Deep Strategy," is the most revolutionary and original of the whole work. He recognized as foremost the potential influence of mechanization, motorization and airpower on the evolution of operational art. Set within a strategy of annihilation, Isserson based a uniquely Soviet style of operational art firmly on the principle of the offense. He believed that only through offensive action could a decisive conclusion be achieved in war. A purely defensive stance would lead only to a reprise of the Great War: protracted struggle and eventual defeat. Isserson envisioned,

That the distribution of an operation in depth would be more fully developed in the western European theater than in ours. Nevertheless, for us *a future operation will no longer be a broken chain of interrupted battles. [Instead] it will be a continuous chain of merged combat efforts throughout the entire depths.* [pages 47-48]

He further clarified and amplified his understanding of the operation by noting,

A modern operation does not constitute a one-act operational effort in a single locale. Modern deep operational deployments require a series of uninterrupted operational efforts that merge into a single whole. In operational terminology, this whole is known as a series of successive operations....*A series of successive operations* **is** *a modern operation.* Without depth, an operation is deprived of its essence and becomes historically conservative, failing to correspond with the new conditions that define it. [page 48]

Isserson goes on to emphasize that,

Under present conditions, we must refer not to a series of successive operations, but to a series of *successive strategic efforts, and to a series of separate campaigns in a single war.* [page 48]

Thus, we can visualize a modern campaign as a *system of consecutive deep operations.* Each of the several constituting operations of a campaign can be viewed as successive "waves" of effort that "ripple" forward into and through the depths of enemy deployments. War itself would then be considered as a *system of consecutive deep campaigns—* air, land and sea—integrated in space and time. Isserson realized as well that, "the blunt facts are that *we are facing a new epoch in military art, and that we have to shift from linear strategy to deep strategy."* [page 48]

The newly emerging technologies of combat aircraft, mechanization and motorization led Isserson to argue that for the first time in history, the offense could become the stronger form of war—a complete rejection of Clausewitz's famous dictum that defense is the stronger form of war. Yet, the Soviet theorist also recognized that the defense could not be completely ignored, that the new military technology would also accrue to the benefit of defenders. Therefore, it was paramount to remember that:

*a modern operation is an operation in depth. It must be
planned for the entire depth, and it must be prepared to
overcome the entire [defensive] depth.* Moreover, it must
be anticipated that the intensity of resistance within this
depth tends to increase and grow denser from front to
rear. [page 55]

The *thickening* of the defense meant that the offensive had to be
structured into more than a single operation. Even under Napoleon, offensive action was guided by the *principle of concentration,* in which all
tactical actions occurred at the same time and place. The new operational
conditions, however, meant that space and time were *detached* from
each other and that not all offensive efforts could occur *at the same time.*
Moreover, because of the increase in defensive depth, these efforts could
not all occur *in the same space.* The concentrated blow of the attacker
was thus *dispersed* into the deep operational space of the defender and
dissipated throughout the temporal fabric of the attacker's maneuver. In
no circumstances under modern conditions could Napoleon's principle of
concentration be achieved.

If the old classical principle of concentration stood repudiated under
the new strategic conditions, what was its new analog? Isserson believed
that,

> *A modern multi-act deep operation cannot be decided
> by a single simultaneous blow of coinciding efforts. It
> requires deep operational reinforcement of these efforts,
> which expand in proximity to the highest point for attain-
> ment of victory.* [page 57]

A defender deeply echeloned in his deployment will likely force the attacker into a commensurate deep echelonment. Each offensive echelon
constitutes an *attacking wave*, while the echelons collectively and suc-
cessively constitute the *operation*. Thus,

> *A modern operation essentially elicits distributed efforts
> in time, thereby conditioning strategy....Modern op-
> erational echelonment is the sequential and continuous
> increase of operational efforts aimed at breaking enemy
> resistance through its whole depth....While deploying
> for a modern deep operation, it is necessary to calculate
> forces and means both along the linear dimension of a
> front and in the new dimension of depth.* [pages 57-58]

From Isserson's writing, a clear visualization quickly becomes apparent: modern warfare resembles the crashing of huge powerful waves against rows of stout seawalls, one behind the other. He held that,

> *Final success will reside with the side having the deeper operational deployments....A front must be broken by means of a decisive operation. A front must be broken and totally crushed throughout its entire depth.* [pages 64-65]

In practice this meant that,

> *A modern deep breakthrough essentially requires two operational assault echelons: an attack echelon for breaking a front tactically; and a breakthrough echelon for inflicting a depth-to-depth blow to shatter and crush enemy resistance through the entire operational depth.* [page 66]

The dynamics of the operational breakthrough have important implications for the geometry of maneuver. The attack echelon operates *concentrically* from exterior lines as a break-in force, while the breakthrough echelon operates *eccentrically* from interior lines to achieve the final breakout. Thus, there emerges a new synthesis and symmetry of modern maneuver, leading finally to the key question: How does one command and control a deep operation?

Under the classical paradigm, control was directed toward the climactic decisive battle with the commander sitting on his horse, surveying the battlefield and flinging dispatch riders hither and yon to control the outcome. With the emergence of the operational paradigm the challenge to command and control became staggering:

> *During the epoch of deep strategy, a deep multi-act, multi-level main battle incorporating all an operation's phenomena will lie from beginning to end within modern operational art's sphere of competence. Otherwise there absolutely cannot be any operational art.* [page 71]

Thus, effective command and control become the sinews of the operation and the new speed-of-light technology, like the radio, becomes the operation's nerves. Technology, structured within new forms of military organization, enables the commander to marshal his massively distributed formations into a coherent deep operation. An appreciation of Isserson leads to the conclusion that the operation becomes the pedestal for the operational artist.

The final part of Isserson's book slips down to the tactical level and addresses contemporary forms of battle. It is the most dated part of the work, though it reinforces the conclusion that the idea of the operation has clearly supplanted the idea of the battle as the intellectual lodestone of modern warfare. In the last section he revisits the technological innovations that transformed tactics, but it is apparent to the reader that Isserson has completed his work. *The Evolution of Operational Art* belongs to posterity.

Conclusion: Distant Hoof Beats

In 2007 Michael Ignatieff wrote, A sense of reality is not just a sense of the world as it is, but as it might be. Like great artists, great [theorists] see possibilities others cannot and then seek to turn them into realties.[12] This assertion is certainly descriptive of G. S. Isserson. He understood that, in order to bring the novel into being, a theorist "needs a sense of timing, of when to leap and when to remain still." And he also understood that theoretical "judgment was the ability to hear, before anyone else, the distant hoof beats of the horse of history."[13] An historian himself, Isserson also comprehended the influence and power of thought on the evolution of Soviet military art. The impact of other thoughtful individuals, like Tukhachevsky, Triandafillov and Svechin, upon a professional institution, largely set free of its intellectual past for one brief but shining moment, was profound. Before most others, these few theorists recognized the military revolution that had taken place. This realization was so rapidly infused into the small body of professional collective wisdom then existing in the Red Army that it constituted a military-intellectual revolution in its own right. That the Soviet Army resisted Stalin's crude attempt to "lobotomize" its brain and survived Hitler's attempt to destroy it stands as enduring testament to the vitality of that revolution. Isserson's *The Evolution of Operational Art* is a legacy for all those who hearken to the hoof beats of tomorrow's conflicts.

James J. Schneider, Ph. D.
Emeritus Professor of Military Theory
School of Advanced Military Studies

Notes

1. James J. Schneider, *The Structure of Strategic Revolution* (Novato, CA: 1994), passim.

2. Jonathan R. Adelman, *Revolutions, Armies and War* (Boulder, CO: Lynne Rienner Publishers, Inc., 1985), *passim.*

3. James J. Schneider, "The Theory of the Empty Battlefield," *Journal of the Royal United Services Institute*, September 1987, 37-44.

4. Walter Klein, *Lost in the Meritocracy* (New York: Doubleday, 2009), 7.

5. See James J. Schneider, *Guerrilla Leader (*New York: Bantam, 2011) for a discussion of heroic and autonomous leadership styles, passim.

6. Richard Simpkin, *Deep Battle* (London: Brassey's Defence Publishers, 1987), 37.

7. M. N. Tukhachevsky, ed., *Future War* (Moscow: Fourth Directorate of the RKKA Staff, 30 June 1928), xi.

8. M. N. Tukhachevsky, *Future War,* 653-54.

9. V. K. Triandafillov, *The Nature of the Operations of Modern Armies* (Ilford, UK: Frank Cass & Co, Ltd, 1994), xxv-xxvii. Triandafillov's influence is so widespread, for example, that his book is still used in the curriculum at the U. S. Army's School of Advanced Military Studies.

10. Refers to pagination in translation of Isserson's text.

11. Unless otherwise indicated, italics are in the original.

12. Michael Ignatieff, "Getting the Iraq War Wrong," *New York Times Magazine,* 5 August 2007, 28.

13. Michael Ignatieff, "Getting the Iraq War Wrong," *New York Times Magazine,* 5 August 2007, 28.

Brigade Commander
Georgii Samoilovich Isserson

THE EVOLUTION
OF
OPERATIONAL ART

Author's Preface to the Second Edition

The second edition of this book appears four years after the first. Naturally, great changes have taken place during this interval. If this work were specifically aimed at formulating an applied theory for conducting modern operations, it would require essential corrections. But, we did not previously posit this aim, nor do we posit it now. This work sets forth the historical and theoretical foundations for new forms of armed combat on an operational scale. The very character of the book makes basic changes unnecessary. The aim of drawing fundamental propositions from conditions of historical development was to foresee the possibilities and conditions for a new era of military art in general.

These propositions have received substantial affirmation from the course of events over the last four years, which have witnessed new and tremendous growth in the armed forces on the European continent. Huge, multimillion-man armies, fully equipped with modern armaments, have no other prospects for use on the contemporary field of titanic battle, except those delineated by the concept of the deep operation.

Whatever the case, this historically-derived concept has never encountered any kind of fundamental objection in principle. Rather, the concept's penetration into the military and theoretical thought of modern official military writers has become all the more evident. In this respect, the military publications of the German fascists are a vivid example. In *Militarwissenschaftliche Rundschau* (no. 2, March 1936), General Ludwig [Beck] writes about commitment in depth into the modern operation by three operational echelons. In the same journal (no. 4, June 1936), General [Waldemar] Erfurth criticizes the old principles of linear deployment. He writes that,

> A disproportionate struggle for width has led to neglecting requirements for echelonment in depth and to a categorical rejection of reserves in the attack.
>
> The World War proved that it was almost impossible either to alter the axis of a main attack or to change a prior decision under conditions of battle across an extended front. The thin lines of the attackers as well as the defenders become firm, immobile and inflexible....
>
> The linear strategy of the recent past must be abandoned in favor of echeloned deployment by powerful operation-

al reserves during both offensive and defensive operations...During a maneuver war of the immediate future, we would conceive of large formations distributed laterally and in depth.

French military thought is no less definite. In an extremely interesting work, *Deux Manoeuvres,* General [Lucien] Loizeau writes about the necessity both to commit large forces at the beginning of an operation and to ensure their continuous insertion from the rear over a protracted period of time. In his opinion, a key requirement is the in-depth echelonment of forces.

The development of operations in depth and the operational depth of battlefields are becoming more and more characteristic of modern conditions. Everything testifies to the fact that one will be severely punished for neglecting these historically-informed perspectives. Our epoch of multimillion-man armies and advanced military technology is an epoch of deep strategy and the deep operation. But one should bear in mind the fact that we are analyzing operations no one has ever conducted. We deal with specific methods of struggle never before tested in combat and operations.

Our research work in the field of operational art is essentially different from similar works of the past, when military scholars like [Alfred von] Schlieffen, [Sigismund Wilhelm von] Schlichting, and [Friedrich von] Bernhardi deduced their operational theories entirely from an analysis of historical experience from recent wars, using well-known and verified data. This historical approach to investigation remains obligatory. It forms the basis of our own work. However, under conditions of the greatest revolutionary era in our construction of socialism, we have managed to create a unique society and army. This fact, together with an unprecedented growth on a daily basis of our productive forces, which yields highly efficient material values by the hour, means that past experience retains for us only the significance that history imparts in a general sense.

We would be powerless to achieve the aims of the present if we failed to go beyond the limits of historical experience, if we failed to reassess it from the perspective of the new conditions of our era, and if we did not mercilessly discard all that was time-worn and stale. We are presently engaged in revolutionary construction, and our operational art acutely perceives this fact. In studying the forms of modern warfare we confront absolutely new tasks that were neither set nor achieved in the past.

There are natural difficulties. Much must be done to delineate precisely and definitively the basic principles for conducting a modern war on an operational scale. This delineation is necessarily determined by the very essence of the deep operation. It is a complicated system that amalgamates all combat efforts into a single centralized and unified complex of actions along a front and in the depths, on land and in the air.

We need to improve our study of the tactics of modern battle, since the outcome of an operation directly depends upon how the enemy is influenced on a tactical scale. Operational forms of warfare mean nothing if they do not involve the crushing power of a direct tactical blow. This is why we decided to publish a second edition of this book, with a new third section that constitutes a separate essay on the historical roots of new forms of battle.

Finally, we think it necessary to repeat the conviction that this book should not be considered a direct guide to action. It would be absurd to teach operational art as a kind of ready-made scheme or recipe. The very essence of operational art presupposes freedom of methods and forms which should be carefully chosen each time to fit a concrete situation. All the propositions we advance in the field of modern operational art should be treated as orienting ideas, which find this or that concrete expression only in a given genuine situation.

Therefore, the present work would be of negative value if the ideas it advocates were treated as ready-made schemes. There can be no such schemes in operational art. We aim to show essential distinctions between the conditions of our era with its new forms of the deep operation and the operational art of the past. This is the only significance ascribed to the propositions advanced in the present work.

Moscow. May 1936
G. Isserson

Author's Preface to the First Edition

At major turning points in history, when an old-regime social order is being destroyed during a titanic struggle and a new society is being built, armed combat as a continuation of politics undergoes basic changes. Revolution replaces evolution in military art. This process forces us to redefine and to solve in new ways all the basic questions of organizing and waging the armed struggle of the proletariat. The capacity of Marxist military scientific research offers boundless possibility for reviewing the basic principles of old-regime military art and for solving a myriad of today's new problems. *The Evolution of Operational Art* is an attempt to study the nature of operations in future war.

This new and little-investigated topic is analyzed in historical and theoretical perspective to work out an applied theory of contemporary operational art. The work advances postulations in concrete and calculated formulations. Yet, as a piece of research, it cannot pretend to be a complete and final answer to the problem. On the contrary, it is proposed that the book's broad discussion might lend impulse to advancing our military-theoretical thought in the field of operational art.

Such an outcome would to a great extent answer the purpose of the book.

Moscow
16 October 1932
G. Isserson

Part One
The Operational Heritage of the Past

1. Ways of Development for Our Operational Art

Contemporary operational art as instruction for the conduct of operations faces a number of new problems. Much remains unknown and unresolved in this field. Colossal changes in technology, armaments, and combat formations, as reflected in the evolution of tactics, remain insufficiently embodied in theory at the level of combat actions along an entire armed front. In contrast with the past, a contemporary operation unfolds under absolutely new political conditions, and on an absolutely different material and technological basis. However, this operation still lacks sound theoretical foundations for the organization of combat actions and for the development of their operational forms.

The entire experience of recent wars, so rich with regard to tactics, still conceals the true nature of future operations. This situation is aggravated by the fact that the World War did not yield a single operation which could be considered an operational solution for the attainment of victory. Certain operations that resulted in actual defeat of the enemy, including, for example, [General A. V.] Samsonov's debacle [at Tannenberg], did not play an essential role in the war as a whole. The grand and fierce battles of 1918 failed to resolve the problem of overcoming fronts on an operational scale and became the highest manifestation of the dead end at which military art had arrived during the epoch of imperialism. The World War drew to a close without resolving the difficulties of organizing and conducting offensive operations.

These difficulties stemmed from the enormous defensive power of the entrenched front, from the declining morale of soldiers during the last years of the war, and from the superiority of defensive over offensive means. Other difficulties included the necessity for the massive concentration of suppressive assets and the complexities of organizing and conducting offensive actions. To put it another way, these difficulties became completely localized in the realm of tactics, and they greatly influenced the conduct of all operations in 1918.

As [General Erich von] Ludendorff has said, "Tactics ought to be placed before strategy." And, actually, offensive operations during the World War were not conducted per operational requirements, but in places that tactical conditions dictated. The main effort was not developed along an axis that promised operational results, but in a locale where tactically the front line could easily be broken. The German offensive of March 1918 was a case in point. The nature of positional war predetermined operational consciousness. It was impossible to overcome the new conditions of combat. And, most important, class contradictions were growing in both capitalist blocs.

In order to instill in every new soldier enough strength to overcome the enemy's resistance in an open offensive, it was necessary to awaken the class will of the masses. Class contradictions had to boil over into an open and armed class struggle, and the imperialist war had to be transformed into a civil war. Our civil war of 1918-1921, with its deep, crushing blows that lasted until the enemy's final defeat, initiated a new epoch in the history of military art and sharply changed the entire nature of armed struggle.

[Carl von] Clausewitz wrote the following about the wars of the French Revolution: "Revolutionary wars overthrew all the former social order and chased the enemy from Chalons to Moscow." One need not be Clausewitz to reverse the geographical order of the last words in this quote to read from east to west and to understand the flexible nature of our revolutionary-class war.

However, the operational essence of the currently unfolding new epoch has not been disclosed in full, especially with regard to controlling huge military masses that are well equipped with modern technologies. The significance of changes during the period after the Russian Civil War remains great. They force us to raise the question of the ratio of the qualitative strength of the defensive and offensive in a different way, with a bias for preponderance in favor of the latter. Under these conditions, the problem of overcoming a firepower-intensive front acquires quite a new meaning. It involves the possibility of "rupturing" the front through its entire depth. In fact, all our military thought aims at solving this problem.

Both in capitalist countries and in our country after the World War and the Russian Civil War, the evolution of military art flowed from a different class basis, but the overall evolution was characterized by a search for new tactical forms for the offensive and the application of new technological means for combat. The short period after the World War constituted an entire epoch in the field of military art, during which tactics underwent greater change than during the entire half-century before the World War. The former period was a time when regulations were reviewed and drawn up anew. New tactics were worked out within several years. It is worth noting that over the entire course of the development of military art, tactics have never changed so rapidly.

Prussia entered the wars of 1866 and1870 with the regulations of 1847, altering them only in 1888. The German army entered the World War with regulations promulgated in 1908. Thus, over a long 70-year period, the Germans changed their regulations only twice.

During the course of intensive socialist construction, we issued provisional field regulations in 1925, and they were replaced by permanent field regulations in 1929. Now, we are once again issuing new field regulations. Thus, for the third time within a decade, we find ourselves with new field regulations. This fast tempo for elaborating field regulations, natural during colossal advances in technology, has become a common phenomenon in the development of military art after the World War.

However, these rapid changes largely reflect and determine the development of the art of war in the field of tactics. The problems of combat in general along an armed front and of conducting military actions on an operational scale have been set aside. Only recently have they again attracted the attention of military-scientific research. Still, pertinent literature remains largely concerned with the general questions of waging war within the framework of politics, strategy, and economics. The practical questions of conducting military actions along an armed front and of the techniques for conducting the contemporary operation find only pale reflection in contemporary literature. During the first years after the World War, the Germans went no further in their publications than an analysis of operations in the World War. After creating a rich military theory following the Franco-Prussian War of 1870-71, they are still digesting the teachings of Schlieffen. In this respect, their military writer [General Wilhelm] Groener had some interesting but hardly new ideas.

In France, [Frederic] Culmann's *Strategie* has appeared, and it is apt to be considered the latest word in teaching about operations. However, Culmann did not present a total operational system. He treated only some questions associated with it. And the most important thing about his work was the fact that his perspective on the future only incompletely envisioned the incorporation of everything that was new on an operational scale.

A number of bourgeois military writers try to replace a somewhat scientific theory for the conduct of operations with vague fantasies on perspectives for future war. But these works, reflecting the class character of contradictions within modern capitalism, testify to just how little the problems of contemporary operational art are being explored by scientific theory.

Only recently, when the ascent of German fascism to power created a war threat of unprecedented scope, has there appeared a number of new works on the nature of contemporary armed conflict.

Much is written about future war. Military writers like Ludendorff, [J. F. C.] Fuller, [Friedrich] Immanuel, [Horst von] Metzsch, [Edouard] Requin, Rocco Morretta, [Ettore] Bastico, and others attempt to predict the nature of future war, with each having his own point of view. Among the new works are many interesting ideas, but nonetheless, they remain mostly speculation. The main topic is what future war will look like. Least of all do various authors analyze and substantiate concrete forms for operations in contemporary war. A notable exception is the French General Loizeau, who, in his *Deux Manoeuvres,* tries to settle in practice a number of questions about contemporary operations. Nevertheless, on the whole, forecasts about future war in foreign literature do not advance any principally new ideas.

On the basis of the greatest revolutionary construction, our military-scientific thought has developed along its own lines. In an analysis of the forms of contemporary armed struggle, we had to be bold in raising and settling a number of new questions. In this respect our literature has evident advantages. [V. K.] Triandafillov's *The Nature of Operations of Modern Armies* is a work notable among others dedicated to the contemporary operation. The scope and nature of the questions treated amount to an elaboration of an entire operational system, which solves a number of problems in a practical context. But one should bear in mind the fact that before his tragic death Triandafillov had radically changed his views on a number of essential questions. On the basis of our achievements his inquiring mind was looking for new and more far-reaching prospects. A tragic accident did not allow him to elaborate a new system of operational views. Meanwhile life has gone far ahead.

In conclusion, instruction about contemporary operations is insufficiently worked out and remains the least elaborated aspect of military art. The fact that this situation has previously occurred in history can hardly be a consolation.

Under capitalist conditions military theory always lagged behind practice, and in the first instance this fact has been reflected in operational questions. To a great extent tactics amount to practice that can be tested in maneuvers and exercises. During peacetime, the conduct of operations is mostly theory that cannot be tested. It is much easier to apply new means on a limited scale than to organize their mass application. Thus, tactics have repeatedly overtaken operational art. At present, this situation is scarcely acceptable. Absolutely different terms of struggle on the armed front, new human material, and fresh means of combat forcefully require new forms and ways of employment on a massive operational scale, where quantity is transformed into quite a different quality.

Before the epoch of imperialism, when armed forces were of comparatively limited strength (the Prussian army of 1870 numbered 500,000 troops), the questions associated with conducting an operation did not attain the status of an independent theoretical subject, for they were entirely solved within the framework for elaborating a concrete war plan. All the questions that [Field Marshal Helmuth von] Moltke faced while preparing for war in 1870 were reduced to the practical elaboration of deployments against France.

Nowadays, a number of complicating factors have arisen, including mass armies, qualitatively diverse means of combat, highly sophisticated technology, very deep columns, the difficulty of deployment into combat formation, and a complex supporting rear. In consequence, the conduct of an operation produces problems that cannot be solved within the framework of a concrete plan of deployment and that require the working out of a general theoretical foundation.

The operator in his practical work now needs a refined theory for the conduct of operations. Thus, operational art as instruction about operations acquires significance as the most important discipline for practical operational work and for control of large troop formations. The topicality of problems associated with operational art flows from other considerations as well. It is quite evident that considerable changes in technology and tactics give rise to no less considerable changes in the conduct of operations. Clausewitz has written, "Changes in the nature of tactics must also influence strategy. If tactical manifestations in a given instance are of a different nature than in another, then strategic manifestations must also change; otherwise, they would not be sequential and rational."

This apparent internal logic has not always been understood. During the era of Moltke, under conditions of new weaponry and modified tactics, everyone still approached battle from the perspective of Napoleonic military art. Within this context, Moltke was a great reformer, for he managed to understand the new conditions and requirements of his time. However, in 1914 the forms and methods for conducting operations differed little from those of Moltke's era. All the factors of armed conflict had undergone qualitative and quantitative change. However, operational control of these factors never underwent any kind of qualitative improvement. Even now, if we give any thought to devising an operation as it is presently envisioned, we would scarcely find any essential changes. Corps are deployed along a single line, sectors for the attack are designated, and missions are assigned in accordance with boundaries...but all this was also done in 1914, and if we go farther back, it was done even in Moltke's era!

Operational art seems intolerably conservative. Meanwhile, present conditions and those of 1914, to say nothing of the conditions during Moltke's era, are completely incompatible. A whole range of primary factors within armed conflict has changed. New armaments, new tactics, and a new kind of a soldier inevitably bring radical and essential changes to the conduct of operations. It is quite clear that modifying factory equipment and putting new machines into operation are factors that basically change the entire process for production and its organization. In the corresponding military sphere, analogous factors would naturally ordain a different organizational configuration of military units. The conduct of contemporary operations should be thoroughly analyzed from this perspective.

But, a consideration of only the new human and material elements would still be insufficient. An operation is a weapon of strategy, while strategy is a weapon of politics. This is why an operation is not the highest stage of armed conflict. Rather, an operation is itself an element within the larger equation, subordinate to war in general.

Based on Clausewitz, comrade [V. I.] Lenin wrote, "only the smallest part of new phenomena in the field of military art can be treated as fresh [military] ideas and inventions, since most of these phenomena stem from new social relations and new social conditions. (*Leninskii sbornik, XII,* 421.) Several factors, including completely changed social conditions, a new social and political life, a different economy, and the new revolutionary and class character of our future war, alter the nature of the operation itself. We occupy a more advantageous position in defining this nature. Marxist-Leninist teachings about war fully clarify the nature of armed struggle. A number of Communist Party documents and Comintern resolutions specify this teaching in the best possible way with regard to the question of future war.

Resolutions of the VI Congress of the Comintern assert, "The coming world war will be not only a mechanized war employing a huge amount of material resources. It will also be a war that involves multimillion-man masses and the majority of the population of the belligerent countries." This is the way the Comintern Congress resolved one of the most essential questions about the relative weight between technology and the masses in future war, and consequently, in the operations of such a war. Only on the basis of Marxist-Leninist teachings about war can we construct a theory of operational art.

In sum, a number of qualitatively new factors, including new social and political conditions, a different range of technological means for combat, new tactical forms for battle, and finally, the great urgency and practical significance of a theory for the conduct of operations, define the basis for the development of our operational art. But we should bear in mind that operational art as instruction about the conduct of operations is an extraordinarily young discipline. In essence, it traces its roots only to the period following the World War, when it first occupied an independent place among military disciplines.

Before the World War, military art admitted only two main elements: strategy as teaching on war, and tactics as teaching on battle. This bifurcated understanding only demonstrated once again how far military theory lagged behind practice.

Even in the second half of the nineteenth century, the evolution of the forms for armed combat exceeded the bounds of this understanding of strategy and tactics. Armed conflict gave birth to a whole chain of combat actions that stretched across a front line and were distributed in depth. These actions exceeded the limits of battle and therefore could not be subsumed into tactics. Because these actions did not embrace the phenomenon of war as a whole, they could not be treated as the teaching of strategy on war. Thus, in theory there opened a considerable gap between strategy and tactics, and this gap in the practice of armed combat was filled by real phenomena of great scope and content. These phenomena required a new understanding which emerged only after the World War under the rubric of operational art as instruction on operations. Consequently, operational art came to occupy an independent place in the now-tripartite division of military art into strategy as teaching on war, operational art as teaching on operations, and tactics as teaching on battle.

However, after having recently become an independent discipline, operational art now faces the task of fundamentally reviewing everything taught about the conduct of operations. Such is quite typical in the history of military art: something new and even recently born suddenly turns out to have aged. Our operational thought cannot fixate on the experience of the World War. This exhausting system of attrition battles, which failed to solve the problem of operationally breaching a front, and whose very slow offensive tempo, requiring four months during 1918 for the allies to push the Germans back only 100 kilometers, cannot become the sole point of departure for developing our theory on the conduct of operations.

Bearing in mind the revolutionary-class character of our future war as a decisive confrontation between two incompatible worlds, we must go farther and demand more from our military theory. The emerging epoch of proletarian revolutions, together with the building of socialism and revolutionary-class wars, undoubtedly foreordains the advent of a new era in military art. As [Friedrich] Engels has said, "the actual liberation of the proletariat, the complete removal of all class distinctions, and the full ownership of the means of production...presuppose creating a new means of waging war.(*Sobranie sochinenii K. Marksa i F. Engel'sa, XIII*, 491-93.) Our operational doctrine faces great challenges which never were and never could have been resolved by the imperialist war [First World War]. These include: breaching a front, waging a deep offensive to pierce and shatter a firepower-intensive front through its entire operational depth, and finally, inflicting lethal, crushing blows aimed at the complete destruction of the enemy. Under these conditions, the basic mission of our operational art is *substantiation and elaboration of the theory of a deep operation for annihilation*.

2. The Evolution of Operational Art before the World War

Working out a theory of operational art is very complicated because of the various roads one must travel. Schlichting wrote that "a new strategic method has never sprung forth like Minerva from Jupiter's head [because] it arises from the peculiarities of an epoch and from corresponding combat means." All the peculiarities of the present-day in their socio-political, economic, military, and industrial dimensions afford material for a definition of operations in future war. But these peculiarities cannot be construed as something permanent. Their developmental tendencies are essential in determining the nature of armed conflict, and they can be traced and perceived only within the dynamic context of the historical process.

To understand the specific nature of the contemporary operation, one must establish the prerequisites and conditions which have caused its birth and determined its evolution over time. This historical approach also reveals the prerequisites that determine the further evolution of operational forms during armed conflict. In historical context, the phenomenon presently known as an operation vividly reveals the characteristic features that have defined the evolution of its nature.

The conduct of war in the era of Napoleon schematically consisted of two basic stages that were far from equal in scope and duration. These stages included a long march along an extended operational line and a short battle in one locale upon completion of the march. Clausewitz described the situation as follows: "In the eyes of strategy, the field of battle is no more than a single point, just as the duration of battle is no more than a single moment." Indeed, when compared with lengthy operational lines, battle during the Napoleonic era was no more than a single point in space and a single moment in time.

This epoch in military art deserves to be called the epoch of the strategy of the single point, for the main task of a commander was to concentrate all his forces at the right time and right place to engage them in a battle that amounted to a one-act tactical phenomenon.

Further, this scheme for military art during the Napoleonic era reflected its own material prerequisites. At the time, firepower was inefficient and insufficient, so its proportional weight was minor. The key factor for producing a telling effect on the enemy was direct shock action by a vital or living force. Before reaching the field of battle, execution required deployment of all vital mass in deep shock columns. These had evolved from the time when the French Revolution had given birth to a new type of soldier, who burned with enthusiasm for battle. It was well understood that such concentration of force could be attained only by launching a mass shock strike along interior lines. This blow shattered the linear combat formations from the time of Frederick the Great.

Mass concentration before battle also stemmed from the material means available for war. One important feature of combat conditions during the Napoleonic era was the fact that the range of human vision (normally 3-4 kilometers) far exceeded the range of shoulder weapons (perhaps 200 meters) and artillery (perhaps 1,200 meters). In these circumstances, adversaries might approach the battle field in sight of one another, yet remain unable to bring their firepower to bear upon one another. This fact explains why the Napoleonic era failed to witness the advent of the meeting engagement flowing directly from contact during the approach march. A meeting engagement presupposes that adversaries can subject each other to fire as soon as they catch sight of each other on the approach. Indeed, the limited range of weaponry during the Napoleonic era accounted for the pause between the approach march to the battlefield and battle itself. It was this pause that permitted preliminary deployment into combat formation while entering the battlefield and before the actual unfolding of battle.

In turn, this circumstance determined the most essential feature of Napoleonic military art. This was the fact that battle constituted the final step, the crowning end point of a long operational line. Battle neither proceeded from nor was determined by the operational line. Instead, battle constituted a separate tactical episode. The best testimony to this fact included the Italian campaign that ended with the peculiar battle of Marengo and the campaign of 1812 with its crowning battle at Borodino. And so, battle during the Napoleonic era was a one-act tactical phenomenon. It possessed no spatial dimension because its scale consisted of a single point, and it had no temporal dimension because it was simply a moment in time. Moreover, it had no depth because it took place in a locale, and finally it was played out as a self-standing tactical episode which bore no organic relationship to the approach march as a whole. Under these conditions, the operation as it is currently understood remained unknown to Napoleonic military art. Indeed, in those days the fundamental characteristics of the operation were undoubtedly absent. Combat remained the sphere of competence for tactics only, since tactics constituted teaching about battle.

However, each historical period is pregnant with a new one and displays new rudimentary tendencies and forms. Thus, even in Napoleon's age one can detect the first signs of new forms of armed combat that exceeded the limits of a single battle. These forms were evident at Ulm, Regensburg, Leipzig, and in the events of 1814. In analyzing the events of 1812, Clausewitz wrote, "Gone were the times when on the battlefield one might view an individual action during which victory was attained with a single blow."

Nevertheless, such phenomena were not characteristic of Napoleon's age. The most typical characteristic was the long operational line crowned by a point that constituted an independent tactical episode. In this situation, strategy's main task was to concentrate all forces simultaneously on the same battlefield and then yield its place to tactics when battle was initiated. Clausewitz described the situation in the following terms: "The moment the enemy approaches closely enough to offer general decisive battle, the time for strategy is over, and it can take a rest." This point of view remained influential for a long time, and played a very conservative role even under completely changed conditions, when it contradicted in principle the phenomenon of the operation that was born soon after.

During the second half of the nineteenth century all of the conditions which had defined Napoleonic military art underwent fundamental change. These conditions, which included the blossoming of industrial capitalism, the introduction of universal military service on the basis of bourgeois society's new productive relations, and technological progress based on advanced industry, created new prerequisites for the development of military systems.

The introduction of rapid-firing rifled weapons played a huge role. Armed with Dreyse rifles, a Prussian battalion of Moltke's era could fire 4,000 rounds per minute. True, the range remained limited (300-400 meters), but it quickly rose to 1,000 and 1,300 meters (the French chassepot). Meanwhile, the introduction of Krupp rifled field guns soon increased artillery ranges to 3.5 kilometers. Under these conditions, the proportional weight of firepower in battle increased to such a degree that it became the main *factor of impact* on the enemy, and thus was laid the foundation for the *epoch of destruction by fire*.

But, firepower-based tactics profoundly conflicted with Napoleon's deep columns, which did not permit application of maximum firepower assets, and which at the same time afforded splendid targets. If firepower had become the most essential factor in battle, then the requirements of firepower necessitated deployment of the greatest number of firepower elements laterally along a single line, so that all might be engaged. During the second half of the nineteenth century, tactics evolved to redeploy the deep column across a broader firepower front, gradually producing the extended skirmish line. The concentration of troop masses before battle in deep, closed shock columns gave way to broad linear deployments having a qualitatively new basis for increased firepower. Schlieffen wrote that:

> If they do not consciously want to limit the number of fighting soldiers, then they must unavoidably think about a dispersed order and extending the front. (*Shliffen, O voine budushchego.*)

Still, for quite some time conservative tactics emphasized the dense concentration of masses within narrow sectors. However, Engels would write, "the soldier turned out to be smarter than the general, and by common sense the soldier arrived at the extended line of fire." The tactical implications of this circumstance immediately influenced the nature of armed combat on the whole by promoting the impulse to laterally-extended combat formations. Moltke taught that:

> More is lost in depth than is lost by extending the front [because] two divisions moving 7-10 kilometers from each other can better and more easily render mutual assistance than if one division followed the other.

17

Another highly significant factor of the nineteenth century led to actions that were more broadly distributed laterally. This factor was the railroad, which accelerated the concentration of an army in a theater of military actions. At the same time, the railroad network's configuration facilitated the army's assembly from diverse points on a large scale. The same number of soldiers (300,000) that Napoleon had so easily led and deployed as a single concentrated whole was deployed by Moltke in 1866 against Austria in three separate armies along a 400-kilometer front. Moltke's dispositions had to account for the configuration of the railroad network and for the trace of the Bohemian border. In contrast, Moltke's initial deployment of the Prussian army in 1870 against France occupied a front of roughly 100 kilometers, which, after forward movement, gradually stretched to 150 kilometers. This lateral extension of the front seemed incredible at the time, and Moltke was roundly criticized for it by his rivals. Conservative theory elevated the fundamentals of Napoleon's military art into a canon of eternal principles without regard to the conditions and requirements of a new epoch. Meanwhile, Moltke's adversaries, including the Austrian Benedek and the French marshals Bazaine and MacMahon, still aimed at concentrating their armies in restrictive spaces, each time confronting a more extended Prussian firepower-intensive front.

During the wars of the second half of the nineteenth century, the two epochs of military art and two schools of military thought contended with one another. And naturally, the advantage lay with the one that perceived the conditions of its time. This fact mattered only because the wars that Prussia waged during the second half of the nineteenth century were historically progressive and because the Franco-Prussian War of 1870-1871 was a part of the bourgeois progressive policy (lasting for decades) of German liberation and unification. The defeat and overthrow of Napoleon III accelerated this liberation. (Lenin, "O programme mira," 25 March 1916.)

From this time military art transitioned to the lateral deployment of forces along a single line, and armies began to enter a theater of military actions along an extended linear front. This was the beginning of a new era in the evolution of military art—*the epoch of linear strategy*. It was not the numerical strength of armed forces that led directly to lateral deployments, since the Prussian army of 1866-1870 was not numerically larger than Napoleon's. The impulse came from new material factors—combat means and railroads. New firepower assets constituted the key factor that initiated deployments laterally along a line, with its corresponding linear strategy. This development was the strongest affirmation for Engels' idea that, "nothing depends more on economic development than the army and navy," and that 'armament, composition, organization, tactics, and strategy depend primarily on the level of production attained at a given moment and on the development of the means of communication."

With the advent of the epoch of linear strategy, a series of new phenomena entered into the unfolding of military events within a theater of war, and these phenomena both exceeded the limits of the single battle field as a point and outgrew the framework of tactics. Once armies began to enter combat across a broad line, combat efforts were distributed across a front, and battle was no longer linked to a single point, but to various points scattered along the front. The main feature of armed conflict during the second half of the nineteenth century was the fact that the single point of Napoleon's era broke down into a series of separate points dispersed in space.

Still, this was not a continuous front. It was a broken front with several separate battle points for application of combat efforts. The Austro-Prussian War of 1866 started with three separate battles (Gitschin, Trautenau and Nachod) spread across a 100-kilometer front. The war of 1870 commenced with two major battles (Spicheren and Worth) taking place simultaneously 60 kilometers from each other. The strategist Moltke confronted the novel problem of combining and directing tactically and spatially independent combat efforts to achieve the general aim of the enemy's defeat. This was the first characteristic sign of the phenomenon known according to current terminology as an operation. And Moltke was at a loss coping with this phenomenon. As Schlichting observed, "The greatest strategist lacked a sufficient understanding of how to combine the actions of separate armies within a theater of war."

In addition to lateral extension across a front, the second half of the nineteenth century witnessed other new combat phenomena. Along with increased frontal width, there appeared the first notable signs of increased depth, and in consequence, changed time. These were changes that the Napoleonic era had not known, for battle in those days had occurred literally in one place and had lasted only a few hours. There were certain objective prerequisites for the appearance of combat actions in a second dimension, that is, in depth. During the second half of the nineteenth century, the increasing range of weaponry soon equaled the range of human vision. It became possible to destroy the enemy by fire the moment he became visible. The range of vision under normal terrain conditions is usually 3-4 kilometers, the same range as the new rifled field guns (3.5 kilometers). The first shots coming from advance guards in sight of each other were immediately followed by others. As Schlieffen wrote, "The moment a bullet left the barrel, it was instantly followed by another." (Shliffen, *O voine budushchego.*) Combat by fire was initiated right from the march and inevitably drew in the rear elements from advancing columns. At the very first shots, the advance guard hastened forward, and no one worried about a pause between the approach march and combat.

This situation created absolutely new conditions for the unfolding of battle. Preliminary concentration before battle, as in Napoleon's time, became impossible. Combat now unfolded right from the march, and this fact accounted for the appearance of the meeting engagement. This phenomenon in its modern sense became possible during the second half of the nineteenth century, when the increased range of weapons equaled the range of vision.

However, this fact was not recognized for a long time: conservative Prussian generals in 1866 responsibly left their artillery with the baggage at the rear of march columns, with the intention to deploy for battle beforehand in accordance with the Napoleonic legacy. But, the actual course of events, conditioned by new armaments, turned out to be stronger than tradition, and the initiative for opening battle passed from the generals to the leading march elements. In addition to the unfolding of battle from the march, combat was no longer localized, but more broadly distributed, thus acquiring the first subtle signs of depth. There was still another important fact: This tactical depth immediately exceeded the limits of battle to display features of operational depth.

During the second half of the nineteenth century, the brief battle of shock action was transformed into a continuous firepower battle that acquired a protracted dimension in time. Battles during Moltke's age extended 10-12 hours. At the same time, they failed to yield the decisive outcome so typical of Napoleon's time. Firepower appeared unable to resolve the issue during one act in a single sector. At the conclusion of a battle, the enemy was not completely destroyed; he gradually retired, reorganized his formations in a new sector, and once again prepared to give battle. Thus, the chain of combat efforts became distributed in depth.

During the war of 1870, three successive main battles occurred in the environs of Metz (Colombey-Nouilly, Mars-la-Tour, and Gravelotte-St. Privat).The whole course of events lasted only six days, during which time the Second Prussian Army completed an approach by its left wing, traversing a distance of 90 kilometers. This interesting set of battles, distributed in depth, possessed every feature of a modern operation. They consisted of separate combat efforts that Moltke combined in space and time for the attainment of an overall general aim. Such was also the nature of the

Sedan march-maneuver, which lasted ten days and which required traversing 150 kilometers through the depths. Thus, during the second half of the nineteenth century, depth became a new combat phenomenon, although it remained rudimentary.

The war of 1870 (before the fall of the Second French Empire) counted only four major elements in depth that constituted separate main battles (Spicheren-Worth, Metz, Sedan, and Paris). These were a chain of separate and mutually independent combat efforts. For the most part they climaxed in a single main battle, the scope of which was reminiscent of great battles during the Napoleonic era. The series of points distributed in space still led to a common point (Koniggratz and Sedan). Strategy's main aim still remained the simultaneous concentration of all available forces in one place. But the difference essentially lay in the fact that concentration proceeded from widespread deployments along a variety of axes that led to concentric envelopment of the enemy.

Characteristic of the epoch of linear strategy was concentric maneuver along exterior converging lines, a development that gave birth to the notion of "Cannae" on an operational scale. But this maneuver from different directions still led to a single main battle. However, the climactic battle of Moltke's era was principally different from that of the previous century. It no longer unfolded as a separate tactical episode, independent of the long operational line. As soon as battle was initiated from the march, with no interval between the two, battle began to flow organically from the march-maneuver, with battle determining organization for movement. The march developed right into combat, and the march-maneuver naturally grew into battle. The plan for the latter was determined by the arrangement of the former.

In 1866, the same Prussian corps, which were initially deployed across a frontage of 400 kilometers, accomplished their envelopment at Koniggratz by closing to within 4-5 kilometers of each other. Under those circumstances, the plan for deployment envisaged the scheme of forthcoming actions. And, since the possibility for altering the initial grouping of forces was limited, the line of deployed corps could not be essentially changed during the course of the offensive.

Napoleon could arrange his march irrespective of future battle, since he had the opportunity to adopt the appropriate combat formation before entry into battle. In contrast, Moltke had to base his deployment and march-maneuver on a definite plan for defeating the enemy. In his era, arrangement of combat actions required perspective and foresight in anticipation of main battle, and this feature has become characteristic of modern operations. Indeed, Moltke had to fashion a perspective that extended from deployment right up to the major battle inclusively.

As for the art of leadership during Moltke's era, there was no boundary between march and combat, between march-maneuver and main battle, between strategy as tactics within the theater of military actions and tactics as the conduct of battle. The command of armies within a theater of military actions had as its purpose main battle, that is, a sphere of competence that lies within contemporary operational art. In addition, a specific distinguishing feature of the strategy of Napoleon's era, the pause before the unfolding of battle, disappeared, having become anomalous under new conditions for the command of armies. This fact long remained vague. The fundamentals of Napoleonic military art persisted, elevated to the status of eternal principles. On the eve of the battle at Sedan, Moltke lost control over his armies, and it was thanks to the initiative of his subordinate commanders that the march-maneuver ended in decisive battle.

Hampered by conservative military theory, the new phenomena and new conditions during the second half of the nineteenth century required some time to penetrate into the realm of cognition. Even at the beginning of the twentieth century, [the Russian strategist G. A.] Leer elaborated his dogmatic system of strategy on the basis of Napoleonic military art. Meanwhile, already during the wars of 1866 and 1870 combat actions had revealed their new character: They were dispersed laterally along a front, they were distributed in depth, and they flowed organically from deployment as a whole. That is, they acquired the most essential features that define an operation. The wars during the second half of the nineteenth century were the historical starting point from which the operation to come into being and on which its evolution would be based.

3. The Evolution of Operational Art in the World War.

The epoch of imperialism offered vast opportunities for the development of the main features of an operation—its lateral dispersion and distribution in depth. The imperialist economy, with its struggle for markets, resources, and capital investment spheres, made war for the partition of the world an inevitable result of the policies of the ruling classes and gave rise to a colossal growth in armaments and in the size of armies. The expansion of the whole military system conditioned the further evolution of the art of war at the turn of the twentieth century and also determined new requirements for war.

On the basis of experience during the war of 1870, Prussian military doctrine concluded that increased firepower precluded the frontal attack. After reviewing 1870, Schlichting wrote that, "attempts to accomplish purely tactical breakthroughs would be practically impossible in the future." Such conclusions were based on results from the battle at Gravelotte-St. Privat, the first example of an attack against a fortified, firepower-intensive front. The attack had assumed a wild, uncontrolled character. It was clear even then that destructive firepower means were incomparably mightier in the defense than on the offense. The attackers suffered enormous losses, and their attacks were unsuccessful. Meanwhile, the defense collapsed after the chance appearance of a small group of attackers on its flank, a phenomenon that immediately gave rise to a repudiation of the frontal attack. It was recognized as impossible because it was considered unnecessary.

There was still much free space for maneuver, and any position could be enveloped. But this fact was not well understood during the era of Moltke. Schlieffen would write, "It was only late at night that a division was directed more by chance than by plan to the enemy's flank and rear, thus teaching commanders unconsciously how to capture strong positions the way it had been done since Leonidas' time."

The whole evolution of military art after the war of 1870 could be characterized as the transfer of combat decision from the front to the flank. This understanding became the basis for Schlieffen's teaching. Thus, linear strategy strove even more to expand the front laterally. Schlieffen wrote: "It is the extended front that decides everything, facilitating envelopment and naturally presupposing a strong and numerous army....Modern battle comes down to the question of struggle for the flanks. The winner will be the one who deploys his reserves not behind the center, but on the extreme flank." Such was the way that military art evolved at the turn of the twentieth century. In turn, the impulse to extend the flanks and to extend the front required an increase in the size of armies. Their growth was well ensured during the epoch of imperialism. By 1914, the Germans maintained an army of two million, a number surpassing that of 1870 by four times. The essence of competition among capitalist military systems before the war of 1914 lay in the greatest possible extension of flanks in order to achieve an enveloping position.

This was a golden age for linear strategy, and this strategy would lead to the continuous linear front. Moreover, the technological evolution of combat means continued. The new quality of firepower weaponry added

still more pressure to combat actions. Even in Moltke's era firepower assets could destroy the enemy as soon as he became visible.

During the period 1870-1914, the range of weaponry failed to make
any appreciable advances. The range of infantry armament increased from
1,200 to 2,000-2,500 meters, which meant that it remained on practically
the same level. The range of light field artillery increased from 3.5 to 5-6
kilometers, which also amounted to little practical change. Although the
range of heavy artillery increased to 11 kilometers, its small numbers had
little influence on increasing combat ranges.

At the turn of the twentieth century, the evolution of weaponry turned
mostly on increased rates of fire. Enormous results were achieved in this
sphere. Increased rates of fire per minute are revealed as follows:

	1870	1914	Remarks
Rifle	5	12/10	Numerator is the rate of fire in theory.
Machine Gun	0	500/250	
Field Gun	2	20/12	Denominator is the rate of fire in practice.

Figure 1. Increased Rates of Fire per Minute.
Source: Original to Author.

The entire linear front became a mechanism for the highly efficient
delivery of continuous firepower. This development represented the flowering of the era of destruction by firepower, a process set in motion during the second half of the nineteenth century with the invention of rifled
weaponry. It was evident that events of great scope and pressure would
transpire. These events would make radical changes in all the conditions
for armed combat and would enter into even greater conflict with a conservative military theory that was so deeply rooted in the Napoleonic
era. Engels wrote: "The full transformation of the entire military system
caused both by the conscription of everyone fit for military service into

multimillion-man armies and by the introduction of weaponry of unprecedented firepower, decisively ended the Bonaparte period of war, making impossible any other war but a world war of unprecedented cruelty and outcomes."

This assertion was immediately proven by events during the imperialist era, which testified to the increased scope of armed combat. During the Russo-Japanese War, the battle of Mukden unfolded across a 150-kilometer front and lasted for three weeks. The main features of an operation, namely the distribution of combat efforts in space and time, grew considerably in a quantitative sense. In 1914, German armies deployed against France across a 340-kilometer front. Such was the linear breadth of the deployed corps that they transitioned into the attack and fought across a frontage of 250 kilometers at the battle of the Marne. And the character of the front was essentially different from the disrupted chain of separate and dispersed points in space that was typical of Moltke's era. In 1914, the continuous front amounted to a single line of points.

The evolving extension of an operation laterally along a front was over. During Moltke's era, the single point of Napoleon's age had proliferated into a number of separate points, and in the twentieth century the separate points became a continuous line. The main question now was just how far the line might stretch.

In addition, concentric maneuver along various separate axes, which was predicated on freedom of maneuver in space, was unsuited to the inflexible front of the twentieth century. The actions of this front occupied all available space in its area of operations. Concentric maneuver was replaced by the wheeling of the entire front along exterior lines, a development that became characteristic of the era. Concentric maneuver along separate axes could now be accomplished only in isolated theaters of war, which left sufficient space for freedom of maneuver. Such maneuver found application in East Prussia during the initial period of war in 1914.

The continuous front in space prompted further evolution of the second feature of an operation, its distribution in depth. This development did not involve extending the lines of operation. In 1914, during the Marne offensive, the operational lines were 400 kilometers in length. Marches during previous centuries had covered similar distances. What was qualitatively different in 1914 was the fact that this entire 400 kilometers of depth was filled with a single chain of combat efforts linked by the general intent of the operational plan. These efforts represented the phases of a single operation or a series of linked successive operations, each of which flowed from the previous one and led to the next. Thus, in the war of 1914, the operational depth acquired the qualitatively new character of a single chain of inter-related combat events. However, this chain in depth was not yet continuous. Battles did not occur throughout the entire depth, but unfolded only in separate sectors.

For example, combat actions filled only 23 percent of the time required by the Marne march-maneuver. This index was even lower in the eastern theater of war. There, battles during August 1914 averaged 20.7 percent of the entire time span, while in September the same figure was 5.5 percent. Therefore, operational combat's intensity in depth remained limited to a discontinuous series of battles. What was qualitatively new was the fact that this discontinuous series constituted a single operational chain.

So it was at the beginning of the twentieth century that an operation *took the shape of a chain of combat efforts along a continuous front, linked in depth, and united by the general intent of defeating or resisting the enemy.*

The challenge to operational art as instruction about the conduct of operations was how to link separate, tactically independent combat efforts in space along a front and in time, i.e. throughout the depths, in order to achieve the general aim. In other words, the challenge was to make the chain of combat efforts a highly efficient system coordinated purposefully and sequentially along the front and throughout the depths to bring about the enemy's defeat. For operational art, the solution for this problem involved contending with the new and complex problem of controlling armies deployed as a continuous front along a single line.

There were earlier portents that the conditions of imperialist war, in which two belligerent coalitions pursued equally aggressive aims, would produce cruel and exhausting conflict. And if the contending parties enjoyed economic parity, the struggle would acquire the form of attrition war. A large number of objective prerequisites, including mass armies, colossal firepower, imperialist war aims that were foreign and hostile to the fighting masses, pointed to such prospects.

As early as 1887, Engels wrote the following about a future war:

> This would be a world war of unprecedented scope and intensity. Eight or nine million soldiers will smother each other, while eating Europe out of house and home like swarms of locusts. The devastation and ruin of a 30-year war will be compressed into 3-4 years, embracing the entire continent. The armies and masses will go wild from

extreme scarcities. Meanwhile, desperate chaos within our artificial mechanism for trade, industry, and credit will lead to universal bankruptcy, the break-up of old states, and their routine ways. The resulting crash will send crowns rolling down brick roads, with no volunteers to pick them up. There will be no way to predict either the outcome or the winner, but one thing will be absolutely evident: total exhaustion and creation of conditions for the final victory of the working class.

These brilliant words by Engels predicted the nature of imperialist armed struggle 30 years in the future.

The final settlement of the larger outcome by this or that means for conducting operations could not be achieved within the framework of imperialist war. This was because operational art during the era of imperialism was practiced by representatives of the old dying class, who failed to comprehend the new requirements of the age and who adhered instead to conservative military theory that was deeply rooted in the age of Napoleon. The actual course of armed conflict soon clearly exposed all the contradictions, by revealing a general inability to achieve even the operational results that were objectively possible. None of the new factors in the evolution of the nature of operations was taken into account. To elaborate a system for contemporary operational art, it is vitally important that these factors be fully disclosed.

During Moltke's era, when armies still did not constitute a continuous front, sufficient freedom for maneuver permitted their shift to the left or right, their deployment along either a single or diverging axes, or even their complete reversal of direction. Under these conditions, the strategist Moltke enjoyed a vast field for operational intervention, which required active operational control during the very course of events. In a situation in which deployed armies formed a single line and occupied the entire expanse of their deployments, maneuver in space by a front to change direction became qualitatively different. A very high art of troop direction was required to benefit from each concrete situation for defeating the enemy. Such direction proceeded from controlling armies firmly, by holding some in place and by advancing others with the aim of wheeling the latter around to the flank and rear. Thus, Schlieffen had said that modern armies ought to be controlled like battalions.

The epoch of linear strategy and the continuous front did not preclude operational maneuver. Broken lines that appeared during the course of combat actions offered enough possibilities for maneuver. But the operational art of command failed to meet the new requirements. Fed on the basics of the Napoleonic school, but now in an era of deployed continuous fronts, this art went on assuming that the conduct of main battle lay beyond its competence, and that it could rest at the onset of main battle. This approach to operational art limited its sphere to the prior grouping of forces and their assignment to definite axes.

This approach left as unperceived indicators within the nature of evolving operations which had demonstrated as early as 1870 that the march-maneuver was now flowing organically into main battle and that under conditions of a continuous front each main battle contained prerequisites for the next operation. This failure to perceive explains why it was not clear under new conditions that operational art required active and continuous control over the course of an entire operation up to the main battle inclusively. Further, this failure to perceive stemmed from the deep-rooted conservatism of military theory, which on the vital question of control got hung up at the level of the early nineteenth century.

In 1914, the conduct of operations boiled down to defining and aiming forces. Right from the beginning armies received their reference points and rushed to them along specific axes. This is what Bernhardi meant before the war, when he compared modern armies with arrows shot from a bow. But arrows in flight are understood to be beyond any further control, and this is what happened to the German armies in 1914. Army formations set off in specific directions, aiming for remote reference points without regard to possible situations which might arise during the offensive and which might require a completely different decision.

In failing to meet new conditions for the conduct of operations, operational art during the World War gave birth to a strategy of the remote aim. Its most characteristic feature was the complete neglect of information about the immediate situation. Analysis of a concrete situation at the beginning of a battle and its utilization afterward were of no operational concern at a given stage in the unfolding of events. Yet, Moltke had taught that,

> Each battle is a stage on the way to new strategic decisions....Depending upon its outcome, the material and moral consequences of battle are so enormous that a new situation is created. It turns out that many of the things that were previously planned can no longer be realized,

while other things, which were thought impossible, now become feasible.

This obvious proposition was forgotten.

Irrespective of the outcome of battle, and irrespective of its locale, armies simply moved along fixed axes. Events took their own objective course, outside any kind of influence from the high command, which learned *ex post facto* about the outcome of battles, with many intervening consequences having set in by that time. Operational art detached itself from the active control of the events, letting them develop along their own fixed axes.

The operation became uncontrolled. This fact became the main contradiction within operational art, and in 1914 operational art was absent from the system for conducting combat actions. In accordance with Napoleonic tradition, main battle was excluded from the sphere of competence for operational art, leaving it useless and resting during the entire Marne march-maneuver. With no place to go and nothing to do, German Headquarters remained hidden deep in the rear, and if it had not existed at all, its absence would have done little to alter the historical course of events.

The result was a whole series of favorable operational situations which would have ensured German success, but that were lost. Thus, during the frontier battles, the Fifth French Army, sandwiched between the Sambre and Meuse, managed to escape imminent destruction. German Headquarters did not even bother itself with an operational analysis of the outcome of the frontier battles. On 27 August 1914, with no reference to the evolving situation, Headquarters drafted the following instructions: "The German armies are ordered to advance in the direction of Paris." The advance became a general offensive, but it was a remote-aim offensive that either ignored the concrete situation or skipped over it. With no relation to the enemy grouping, the offensive became groundless and purposeless. In fact, the affair simply amounted to the mechanical transfer of an unchanged grouping into the depths between the Rhine and the Marne.

It seemed sufficient to press operational efforts forward, as if such were the very essence of an operation. The actual aim of defeating the enemy's vital force dropped from operational view. Operational art was least of all concerned with settling the question of where and how to destroy the enemy. This question was replaced by the question of when to reach a place. The attackers were only pushing the enemy back along the entire front instead of coming to grips with, defeating, and destroying him.

Indeed, in this situation defeat of the enemy's vital forces was impossible. An offensive operation had been transformed into an expulsion operation.

Thus, linear strategy was deprived of its essence, the intent to shock and destroy, on which it had been based from the time of its birth during the second half of the nineteenth century. This development was the first indicator of linear strategy's degradation. This development also testified to the complete failure of operational art during the epoch of imperialism to meet new requirements for the control of armies in the twentieth century.

The conservative influence of false methods for operational control was so deeply rooted that the "arrow-like-attack" found full repetition in quite a different army under the quite different conditions of a revolutionary-class war. During 1920, our march to the Vistula once again displayed features of linear strategy on a wide scale. This march was analogous to that of the Germans to the Marne, in so far as methods of operational control were concerned. Once again armies of specific groupings were assigned specific directions of advance to the immense depth of 600 kilometers. Once again, they received remote reference points, and once again the immediate situation was ignored. And, once again a sweeping straightforward advance was made, irrespective of the concrete situation. Finally, once again operational command was isolated from the course of events, while "resting" deep in the rear.

The result was a number of brilliant opportunities lost. On the Neman, Narew, and Wkra, the Third Cavalry Corps and the Fourth Army occupied advanced positions, but they made no use of the immediate situation. Instead of turning to attack the enemy's flank and rear for operational benefit, they headed each time for distant reference points, like arrows shot past the open Polish flank. The outcome was the mechanical transfer of an unchanged grouping from the Dvina to the Vistula during a clearly useless "general offensive." Comrade [I. V.] Stalin concluded that: "A sweeping advance means death for the offensive." (Stalin, *Politicheskii otchet TsK XVI s"ezdu VKP (b)*.)

Linear strategy was sentenced to death the moment it became simply an advancing wall, before which the retreating enemy might freely regroup to counterattack. And then it turned out that the wall's advance was no more inevitable than its movement backward or its defeat from a simple blow to the flank. This was a result of the grand contradiction within operational efforts focused on the assignment of groupings to a single unchanging axis. When this happened, operational art was at a complete loss because it was unable to comprehend the operational sense of transpiring events.

Now, a [Lieutenant Colonel Richard] Hentsch, as the only embodiment of a system of operational control, must solve that which in fact had become unsolvable.* Any change in the grouping of armies engaged in battle along their whole front could be accomplished only by changing the correlation of forces along specific axes. Such would require reinforcement of the front from the rear and the availability of deep reserves. But linear strategy was called linear strategy because it rejected the very idea of operational reserves. In the spirit of Napoleonic times, main battle was still treated as a one-act effort, requiring the simultaneous engagement of all available forces.

Everyone still quoted Clausewitz: "All the available forces assigned to achieve the strategic aim are to be engaged simultaneously. The effect of their engagement will be greater if everything is compressed into a single moment." But that which the great thinker had correctly concluded from the experience of the Napoleonic wars turned out completely incorrect in the twentieth century. By that time, the operation had come to consist of more than one act; it now consisted of a series of successive combat efforts distributed in depth.

When small reserves were required to parry enemy blows on the Marne and Vistula, the operational leadership did not have a single division at its disposal. On this less than high note linear strategy was practically done for. Meanwhile, if operational art could put itself into a situation in which it was absolutely powerless to do anything, it had died. And then, naturally, past teachings were recalled, and Schlieffen's ghost received a plea for help: It is necessary to look for decision on the flanks; victory belongs to him whose flank is longer. Salvation was sought in such maxims, and so began the frantic rush to the sea. But no one accounted for the fact that the enemy could do the same thing, and consequently the front was only extended laterally.

*Editor's Note: Lieutenant Colonel Richard Hentsch was a general staff officer in the Imperial German Army at the outset of the First World War. During the First Battle of the Marne in September 1914, Lieutenant Colonel Hentsch acted as the Supreme Army Command's plenipotentiary and personally gave orders for a retreat to several field armies at a time when the German offensive was successfully advancing against French and British forces.

At this stage, operational art in the World War met with even graver unsolvable contradictions. Already at the dawn of the twentieth century, one could predict that the growing strength of armies would exceed the physical limits of the frontages they might occupy. Lateral space was objectively limited either by natural conditions or by the geographical location of neighboring countries. The impulse for belligerents to extend their flanks in space would inevitably lead to an encounter with natural barriers. In 1914, such a situation occurred on the western front during the second month of the war. After extending laterally 700 kilometers in space, the flanks of the western front reached their limits. These were the sea in the north and neutral Switzerland in the south. There was no more space to accommodate lateral extension. The distribution of combat efforts along a front, the first characteristic of an operation, was now complete for the World War. The extended front had reached its natural geographical limits, and these could not be overcome without dredging the sea.

Linear strategy thereby came to produce its own antithesis. Its whole essence had been to extend the front laterally for envelopment, thereby avoiding the frontal attack. Now, the possibility for lateral extension was lost, and, along with it, freedom of maneuver along a line. In consequence, linear strategy lost the essence of its reason for being.

Its evolution contained all the factors which led to its self-negation. Its ideologue, Schlieffen, failed to foresee its demise. Inspired by the idea of Cannae, his teaching on strengthening and extending the enveloping flank as the supreme formulation of linear strategy appeared at the very moment when such a strategy was already doomed and when the actual course of events contained all the features of its negation. *Schlieffen wrote Cannae too late;* this outstanding author should have lived earlier.

When fronts confronted each other, linear strategy was in fact already finished. There was no other way out except a resort to the breakthrough. In the World War, things that had been considered impossible after the war of 1870 were now recognized as necessary. The chance division suddenly appearing on the enemy's flank could no longer teach the Germans how to capture a strong position, as had been the case since Leonidas' time. There was no longer any flank. So it was necessary to return to the battle of Gravelotte-St. Privat, to transform a wild and uncontrolled attack into the proper breakthrough of a fortified line. Evolution thus came full circle. And, this evolution led to the grand frontal battles of 1918 and created a new stage in the development of armed conflict. It became evident that the epoch of linear strategy was over, and that solution for the breakthrough problem had to be sought along new lines in the evolution of operational art.

By this time, the imperialist war had fully exposed its protracted and exhausting character. The most essential task was to overcome the firepower-intensive front. The tactical content of this task became an end in itself, and operational art, charged with organizing and supporting the frontal attack, was put in service of tactics. Indeed, the task required superiority of offensive over defensive means. This great technical problem had already acquired urgency at the beginning of the destruction-by-fire epoch. Superiority of defensive over offensive means had already been evident before the World War. This fact had been a prerequisite for shifting the center of gravity from the center to the flanks. Schlieffen was also preoccupied with providing the German army with more powerful offensive means. This preoccupation was expressed concretely in a draft project for the introduction of heavy artillery, which the German army was the first to introduce into its field forces.*

But even with these means it was impossible to solve the problem of the proper correlation of defensive and offensive means in favor of the latter.

Nowhere had the entire offensive to the Marne overcome any firepower-intensive front. The offensive was capable only of pushing the latter back. The operation for destruction became an operation for pushing the enemy back, and this development in itself became another factor in the degeneration of linear strategy. When continuous fronts found their limits, competition between defensive and offensive means became the axis around which the evolving technical approaches to combat turned. This situation has continued until the present day.

In a military-technical perspective, events of the World War after the end of 1914 were very informative with respect to the struggle between offensive and defensive means. At first competition undoubtedly favored the latter. It was much cheaper and easier to mass produce machine guns as a destructive fire means than to produce artillery pieces, which were the main means for neutralizing machine guns. If the number of machine guns per division increased 20 times on the average over the four-year course of the World War, then divisional artillery grew only two-fold. Firepower

---

*Author's Note: It is interesting to note that in Schlieffen's report on the introduction of heavy field artillery there was a strange notation: "Is the Chief of the General Staff willing to make heavy artillery mobile?" To which Schlieffen answered: "Yes, of course."

superiority remained with the defense. Suppression required a huge concentration of artillery assets. The fixed average quota was 60 field guns per kilometer of front. In fact, this quota was considerably exceeded, with up to 100 guns and more per kilometer of front.

However, the concentration of artillery suppressive assets could not in the final analysis overcome the firepower-intensive front. In fact, suppression affected only the forward defensive belt, leaving the defensive depths largely untouched. Offensive artillery was incapable of reaching into the entire tactical depth, for artillery lagged behind the attacking infantry. This was not a question of suppressive firepower, but a question of the mobility of suppressive assets, which confronted battlefield obstacles that were insurmountable for wheeled and horse-drawn conveyances. For attacking infantry, the tragedy lay in the fact that after 3-4 hours of a successful attack, only a small number of the 100 supporting pieces continued firing. At this point, the exhausted attack was suspended.

It became evident that the problem must be solved not only by increasing the number of suppressive assets along the front, but also by inventing new ones. It was necessary to invent a means of firepower suppression, which, first of all, would be protected against fire, that is, armored against bullets. Second, the means had to be mobile in any terrain in order to penetrate the defensive depths, while directly suppressing and destroying at point-blank range the enemy's means of firepower destruction. This situation gave rise to the tank, a combination of the internal combustion engine, track-driven locomotion, armor, and firepower. The very appearance of the tank held great significance for re-establishing the superiority of offensive over defensive means.

The necessity to neutralize the entire tactical defensive depth brought to life other combat means: the aircraft as an airborne conveyance of firepower, and chemical contamination that exceeded conventional trajectories to achieve the immediate envelopment of space. Colossal technological progress during the World War was brought about by the new conditions of trench warfare and ensured by advanced industrial development. These impulses fed the process for solving the problem of regaining the superiority of offensive over defensive means.

However, these developments unfolded more quickly than theory. In practice, at first even attacks involving tanks failed. The reasons included lack of tactical skill in their employment and the restricted scope of their introduction to combat. The tank could not immediately resolve the problem of overcoming a firepower-intensive front. In 1918, the Germans provided the first tactical solution to the problem without tanks. Only at the end of the war did the new offensive means demonstrate the tactical possibility of breaching a front.

However, these solutions appeared under conditions in which the imperialist war was already becoming a civil war in Germany, with the masses turning their weapons against the ruling classes. The question of the power of the defense subsequently assumed a different political content. Nevertheless, the last period of the World War made available the data necessary for a tactical solution to the problem of overcoming the firepower-intensive front. But right at this stage operational art turned out to be absolutely helpless. It became fully engaged in the tactical organization and material support for the breakthrough, and thus practically liquidated itself as the art of conducting an operation. The very idea that tactics should be made superior to strategy testified to the fact that operational art had lost its meaning.

Opposing tactics, operational art came in senseless conflict with it. Really, tactics and operational art are notions from the same category, and they differ only in scope and dynamics. They not only co-exist during combat actions, but they organically flow into one another. If a tactical effort does not give rise to operational success, then the effort becomes in essence a useless fact. A tactical effort is only a step on the way to the aim and can never be an end in itself. The latter characterized the situation in 1918, and the problem even now is often understood in the same way. Thus, [the French Lieutenant Colonel Gaston] Duffour speaks about the experience of 1918 like this: "The continuous fortified front is no more than a simple wall behind which strategic maneuver can unfold. *The front has become the main objective of this maneuver.*" (Iz lektsii, chitannykh vo frantsuzskoi akademii)

The entire problem of a breakthrough was reduced to tactically breaching a front. The question was settled only on a tactical scale. The glorification of tactical efforts as operational successes was complete. Still, [the French General Marie Eugene] Debeney writes, "A characteristic feature of the breakthroughs of 1918 was the fact that they envisioned only the first stage of breaching a front, whereas the development of the operation was not taken into account." (Ibid.)

Operational art did not ensure that the tactical efforts for breaching a front evolved into a total operational breakthrough and defeat of the enemy. This shortcoming constituted the essence of operational art's bankruptcy during the World War. The combining of tactical efforts along a front, a main feature within the conduct of operations, fell out of practice.

Combat efforts occurred outside any system and without any prospect for linking them across space and time in the interests of attaining a common aim. The attacker rushed into fierce battle on this or that sector of the front, at best to hammer a nail in each of them. This mode of action was doomed to failure because it was incapable of promoting the attainment of decisive aims. Instead, it became a system of attrition warfare with limited aims. Even during our epoch, this system is often treated as a historical necessity and as the most viable theory for breaching a fortified front.

During the imperialist World War, this system was grounded in a number of political and economic prerequisites. In the circumstances of 1918, this or that method for conducting a breakthrough operation could not finally decide the outcome of the war, but this or that method could naturally influence the political, economic and military situation of the belligerents. Decision had to be sought along other lines. But the political failure of combat actions did not necessarily presuppose their operational absurdity. The essence of operational art involves accounting not only for objective conditions, but also overcoming and mastering them within the limits of objective possibility.

The system of battles for attrition was incapable of finding an operational solution to the problem of breaching the continuous front, and was therefore senseless. As for exhausting the enemy, the system exhausted the attackers more than the defenders. The whole thing was a senseless system of self-attrition. This fact is vividly evident from a comparison of losses among attacking and defending forces and means during all the breakthroughs of 1916-1918.

The application of this system exposed all the powerlessness of an operational art which had come to a dead end. Operational art had practically been transformed into a senseless system for hammering nails. But walls do not fall as a result of hammering nails into them. To bring down a wall, one must undermine its very foundations and flow through the resulting gaps. However, in this regard, operational art turned out to be even more helpless. The prospects for the operational development of tactical efforts in depth were not at all foreseen. There were no operational echelons available for development of the breakthrough. Their absence resulted from the last influences of a dying linear strategy.

When a gap appeared in the continuous front, as was the case during the German offensive of March 1918, the attackers had no assets to take the blow into the depths, thereby transforming a tactical breach into an operational breakthrough and ultimate defeat of the enemy. All the enormous efforts that went into the tactical organization of the breakthrough, including the entire technological improvement of armaments and the massive concentration of forces and suppressive assets, went for naught. Tactical success did not lead to operational success.

History must have looked with cruel irony on the German command, when, during the battle of Picardy on 25 March 1918, not a single German soldier entered the 15-kilometer gap appearing between the English and French lines. In *Franzoesisch-englische Kritik des Weltkrieges,* General [Hermann von] Kuhl writes, "if cavalry had penetrated the vast gap between the English and French armies, it would have surprised and delayed the shifting of French divisions by automotive and rail transport. Cavalry would have destroyed the approaching unprotected enemy artillery and sown panic and fear in the English and French rear."

But alas, there was no such cavalry, and no one had even thought about taking advantage of tactical success to develop the breakthrough. There was no panic in the English and French rear areas. On the contrary, the timely arrival of reserves quickly sealed the breach, and in historical perspective it remained unclear why the breach had been effected in the first place.

It was senseless to break down a door if there was no one to go through it. This is what the penetrations of 1918 were like. The imperialist war failed to solve the problem of conducting a breakthrough. The war ended without demonstrating the possibility of accomplishing a breakthrough on an operational scale.

Still, the German front fell. However, explanation for this event went well beyond the bounds of purely military causation. German defeat in 1918 was more a function of internal rather than external causes. Much stemmed from the growing revolutionary nature of the masses, a development that led both to the collapse of the fighting front and to the overthrow of the German monarchy. Indeed, this process was facilitated by the Entente's colossal economic advantage in forces and means. But it was not a victory for operational art. Even after the German front had lost its defensive coherence, the allies had to spend four months pushing the defeated Germans back only 100 kilometers. [Marshal Ferdinand] Foch did not even intend to end the war in 1918. He was preparing for a general offensive during the following year. But before his decisive attack could begin, the Germans threw down their weapons on the battlefield, and the result was achieved.

Despite all this, Culmann dared to declare: "During the last four months of the war, the French high command showed how in modern war a breakthrough must be accomplished, as well as the results it can produce." (Kiul'man, *Strategiia. Biblioteka inostrannoi voennoi literatury,* 64.) Such boasting sounds horribly ironic in light of the total embarrassment that overwhelmed the general staffs of the imperialist countries during the last period of the World War.

The operational art of the era turned out to be powerless for solving the new problems inherent in the nature of contemporary armed conflict. Operational art got stuck on the level of linear strategy and became weak when this strategy met its nemesis. The problem of operationally overcoming the firepower-intensive front remained unsolved.

This is the operational result with which we confront the dawning day. Under completely changed conditions, including the new political content of war, a new army, and new material and technological foundations, our operational art must solve a problem which could never be solved under conditions of the imperialist war. Linear strategy started with brilliant decisions during the wars of nationalism in the second half of the nineteenth century. During the global imperialist war of 1914-1918, that strategy reached its own self-negation. And, now, during an era of revolutionary-class wars, a new solution must be found. Herein lies the grand challenge for our operational art.

Part Two
The Foundations of Deep Strategy

1. The Basic Principles of Our Operational Art

Any war should first of all be treated with respect to its possible character and its general features as based on political relations and categories. Politics is the only thing entitled to a superior position for governing the general direction of a war.

—Clausewitz

It is evident that the evolution of our operational art along new paths must first of flow from the nature of our future war as a revolutionary-class war. As the highest manifestation of class contradictions within two competing social systems, this war will assume the nature of a decisive class war with world-wide historical importance. By its very radical nature, this war between nations and classes will reach the very limits of intensity. History clearly shows that wars grow in intensity in accordance with changes in their political character.

The wars of the French Revolution instantly engaged huge masses and attained unprecedented scope. During the second half of the nineteenth century, the intensity of the wars of nationalism seemed very unusual to contemporaries. Although this intensity supported political aims, it was nonetheless historically insignificant. Wars for national unification did not inspire reactionary forces to conduct war in a life-or-death manner to the last ounce of strength. Politics did not demand that the losers cede their state sovereignty. Usually, at the last moment, an agreement was easily reached.*

The World War of 1914-1918 was quite different in intensity. The reactionary imperialist character of a war for division of the world and global hegemony was the continuation of acute economic competition among capitalist countries at the last stage of their development. Therefore, the character of this war advanced aims for the total economic enslavement of the adversary, a feature that made the intensity of this conflict unprecedented in history.

However, the internal contradictions within imperialism led to a situation on the Eastern Front in 1917, in which intensity provoked an intense

*Author's Note: The clearest example was the outcome of the Austro-Prussian War of 1866.

bitterness that awakened proletarian class consciousness. From this development was born the antithesis of the imperialist war: revolution, fraternization, and promotion of international solidarity among the laboring masses. Thus was realized Lenin's great slogan for turning the imperialist war into a civil war.

Meanwhile, the intensity of struggle within wars of national liberation acquired a new revolutionary content. This content evoked a new intensity that resulted from a vast rising of the enslaved masses in their struggle against exploitation.

In the emerging epoch of socialist revolutions and revolutionary-class wars, the complex system of socio-political relations foreordains the inevitability of three types of wars: imperialist, national liberation, and revolutionary-class. All will have specific features and varying intensities. Civil wars of a revolutionary-class character will attain the highest intensity. As manifestations of class struggle at its highest stage of development, these wars will be of unprecedented intensity, because of the antagonisms between opposing classes, the differences between contending socialist and capitalist economic systems, and the decisive aims of overthrowing and excluding the opposition. Revolutionary-class wars represent a concentrated manifestation of conflict at its highest intensity. They will crown the last stage of war as a social phenomenon by destroying the very institution of war itself. These wars will last for a considerable period of time because their mandate is to resolve the great historical problem of transition to a new communist society of free labor. These wars will also embrace a considerable part of the globe. Our civil war of 1918-1921 was only the beginning of these wars and of even greater events to come.

At the VIII Congress of Soviets, comrade Lenin said, "A long series of wars decided the destiny of all revolutions, even the greatest. Our revolution is also great. We have completed the first period of these wars, and we have to prepare for the second." (Leninskii sbornik, XXVI, 35) Waging these wars is not only a question of specific contradictions between belligerents, but also a question of resolving the historical dispute between two incompatible epochs and systems on a global scale. These wars will solve the general historical problem of liberating the multimillion-masses of the enslaved. It is the historical significance of future war that predetermines its decisive character and immense intensity.

If, in 1877, Engels wrote that a future imperialist war would be a "world war of unprecedented scope and intensity," it is difficult to find proper words to characterize the even greater unprecedented scope and intensity of revolutionary-class wars. These wars will involve huge multi-million-man masses.* These wars will be waged on an advanced material-technological basis, which current industrial development enriches with a combat arsenal unique in its killing power. These wars will require colossal material resources and immense economic strain.

The only possible outcome of these wars is the destruction of capitalism and the victory of the new socialist order. Indeed, never before has struggle been waged for the sake of such lofty aims. Not a single army in the world has thus far been destined to solve such grand historical problems.

This is the mission of our Red Army as the first class army of the proletarian dictatorship. The key point of departure, the historical significance of revolutionary-class war with "its essence based on political categories and relations," determines the character of the combat front and the nature of our future operations. [M. V.] Frunze has said the following about this war: "If it is class war, and if it is a civil war, then the only possible outcome will be total defeat of the opposition. Half-measures will be impossible once the war has started." (Sobranie sochinenii, I, 400.) He continued, "Judging by the deep contradictions between the two incompatible systems, it is clear that the upcoming clash, once it begins, will be decisive. Combat will continue to death, until one side emerges victorious." (Ibid., III, 112.)

The decisive character of confrontation determines the decisive nature of military operations. These will not be slow, protracted operations of attrition for limited aims, but will in essence be active crushing blows with decisive aims. The character of operations will also be determined by modern technological means, which are speedy, mobile, and highly efficient in their combat application.

The belligerents' attacking forces will not be equal in these decisive operations. Our side's war against the imperialist aggressors will be historically progressive and consequently just. We will be defending and achieving goals of world-wide historical importance. Already during our civil

---

*Author's Note: If bourgeois military writers ([Georg] Soldan, [Hans von] Seeckt, and Fuller) propagate a theory of small professional armies, then their work reveals irreconcilable contradictions over the development of military systems within capitalist countries. Such armies do not meet actual present-day requirements, since fascism advances a clear-cut program for mass armaments.

war of 1918-21, we emerged as a progressive factor of global significance. Lenin wrote the following about the Polish-Soviet war: "In the summer of 1920 Soviet Russia emerged not only as a force defending itself from the violence of the attacking Polish white guards, but also as a global force capable of breaking the Versailles Treaty and liberating millions of people in the majority of the world's countries." (Leninskii sbornik, XV, 419.)

In future war, this world-wide historical role endows us with great power for destruction of the class enemy. This power flows from the political situation in which a young historically progressive class is pitted against the old decaying world of capitalism. Progressive and just war aims have always endowed revolutionary armies with great attacking power. The actions of the armies of the great French Revolution clearly proved this point. As the army of the greatest socialist revolution, the Red Army already displayed its immense offensive power during the civil war of 1918-1921. History demonstrates that the offensive achieves great historical aims, and that a revolutionary army must always be ready for decisive offensive operations. Already in 1905, Lenin wrote, "The great questions of political freedom and class struggle can be resolved in the final analysis only by force, and we must be concerned not only with the preparation and organization of this force, but also with its active use, both defensively and offensively." (Ibid., VIII, 42) This bequest of the leader remains the main directive for our military system to this very day.

The basic principles of our military preparation, of our operational art, are the principles of the offensive.

Here there is no contradiction with our policy for peace. We have always fought and will always fight against war with all our strength. Our policy of peace is constant. The famous words of comrade Stalin say, "We do not want an inch of foreign soil, but we shall never cede an inch of our soil to anyone." And in this determination to defend the first country of socialism is rooted an immense active force, ready to defeat and destroy any attacking class enemy.

For us, the entire essence of class struggle transforms a progressive war into a strategic offensive with fierce and destructive blows against any enemy attacking us. As a continuation of the civil war of 1918-1921 at the next stage of development, our future war can be grounded only on the principles *of an offensive strategy of annihilation.*

Comrade [Kliment] Voroshilov has said, "We have to arrange matters so that we gain victory in future war with little blood-letting. We will wage this future war on the territory of that country which first draws the sword against us." (From a speech at the IX All-Union Komsomol Congress.) This idea determines the basic principles of our operational doctrine for the decisive offensive as specified in the 1936 Field Regulation. It reads that, "Any attack on the workers' and peasants' socialist state will be repulsed by carrying military actions into the territory of the attacker with all the might of the armed forces of the Soviet Union. The Red Army will conduct combat actions for annihilation."

This guiding article within the Field Regulation affords the basis for elaborating the theory of our operation art as the art of waging destructive offensive operations with the decisive aim of completely overthrowing the enemy. The challenge for our operational art is to create a new and brilliant model for military art in an absolutely new historical situation, with a new army, with a new material and technological basis, and with new content and forms. The great aims of war cannot but elicit equally great operational acts. Never has the strategy of destruction been so well grounded historically. And, never has it had enjoyed such favorable prerequisites for implementation.

2. The Evolving Nature of Operations in Future War

The basis for our theory of operational art is the concept of the most decisive offensive operation. The whole nature of future war testifies to the grand scope of this operation, thereby determining the further evolution of its main features. The historical character of operations has evolved along two main lines: lateral extension across a front, and distribution in depth. The development of the first feature, lateral extension across a front, reached its apogee during the World War of 1914-18. Armed combat filled an entire continuous front to merge combat efforts into a single line that was extended laterally to its full geographical limits.

We have no reason to assume that future war will reverse the evolution of this feature. We cannot be party to the contradictions inherent in bourgeois theories about small professional armies. In the opinion of their supporters, these theories would reverse the development of the above-mentioned feature and re-introduce the interrupted front with separate points for the application of combat efforts in space. The course of history cannot be reversed, and we must assume the opposite. That is, we must assume that operations in future war will all the more proliferate along extended lateral fronts, as long as geographical conditions permit.

Our western border alone stretches 3,000 kilometers from the Arctic Ocean to the Black Sea. The entire extent is vulnerable to intervention. This problem involves not only our western border, because our Far Eastern borders are also vulnerable. Indeed, never before has our strategy faced such a vast scope for confronting potential continuous combat fronts. Under these strategic conditions, there can be no talk about the degradation of a front's possibility for lateral extension.

While assessing this question on an operational scale, we should bear in mind that we can expect to confront on the average one division for every 10-12 kilometers of frontage along the 800-kilometer Soviet-Polish border. And, since this strategic front becomes narrower as one advances from east to west, we might confront one division over every 6-8 kilometers after reaching the Vistula-San meridian. Moreover, one should take into account the fact that additional troop mobilizations will increase operational densities. Naturally, tactical densities might be considerably higher.

However, unevenness in front-line densities can result from the creation of strong shock groupings that leave gaps or weakened sectors along the front. This circumstance, along with the immense length of our western border, forces us to assume the existence of operational windows within our western theater, even against the general strategic background of a continuous front. Operational flanks might still be found. Or, alternatively, modern mobile and high speed combat assets (motorized and mechanized units, cavalry, and aviation), if properly employed, can create such circumstances. This possibility ought to be foreseen in basic operational task assignments during the initial period of war.

Our strategic conditions share some features with those of the Franco-German front at the beginning of the World War in 1914, when German armies on the right wing still retained freedom of maneuver during the approach march. For us, these conditions mean that the prerequisites for linear strategy have not completely vanished. Meanwhile, within our separate eastern theaters of military actions, linear strategy will still find full application. There are no pronounced borders between historical epochs. Having created prerequisites for new conditions, the historical process still retains features of the old, and transition from old to new occurs according to the dynamics of dialectical development.

Therefore, we assume the possibility of enveloping maneuvers along exterior lines during the initial period of war. The idea that our front would directly confront an enemy front during the very first days of a future war would, of course, amount to a mechanical transfer of the conditions of the Franco-German front to our theater of military actions. On the Franco-German front, the preconditions for a linear strategy of envelopment were basically non-existent. In larger perspective, however, we must foresee the inevitability, or at least the probability, of the appearance of frontal confrontation more rapidly and on even firmer grounds than was the case with the Franco-German front during the beginning of the war in 1914.

Clear foresight of this phenomenon is predicated on the entire historical evolution of the nature of the operation. The central challenge for our operational art is to be ready in all respects for the dialectical transition from enveloping linear maneuver to the deep frontal penetration. This necessity flows immediately from the requirement for transition from one operational method to another. Extraordinarily weighty considerations force us to fashion just this operational forecast.

There is a certain internal logic within contemporary strategic deployments. *Contemporary deployments do not tolerate gaps.* Deployments occupy almost the entire expanse of a front.

Belligerents look for flanks and the possibility for envelopment, while each fears probable envelopment of an uncovered flank. Therefore, deployments aim to cover the entire expanse of a front. A front, consequently, tends to maximum lateral extension. In the end, when all forces are deployed in a theater of military actions, gaps may no longer exist. Under modern conditions, weak forces mean only a weakly-occupied front. Still, it is a front and not merely the deployment of separate groups with gaps in-between. At present, even poorly occupied fronts rest on defensive lines, thus attaining a certain power for resistance.

It is quite evident that in contemporary circumstances defense requires that everything should be done to construct a fortified front. Modern means, including obstacles, chemical assets, the mechanization of labor, and fast-hardening concrete, provide more possibilities for fortification than ever before. The fortified front appeared during the World War as a result of linear operations and the absence of a shock penetrating force. At present, in many cases, the fortified line is prepared beforehand. It predetermines the character of operations, preceding their initiation and determining their course. The continuous fortified Maginot Line along the Franco-German border affords excellent proof of this phenomenon.

After the World War, [Marshal Ferdinand] Foch wrote, "A nation which enters a war in hopes of hiding behind fortified trenches while its armies deploy is facing catastrophe." Nevertheless, many look to avoid catastrophe in fortified trenches.

In the end, a confrontation between fronts cannot be excluded in many cases right from the beginning of a war. Emerging opportunities for operational maneuver permit the extension of one's flank with fresh forces more rapidly than before, thereby opposing an enveloping enemy with a new grouping. During the German offensive to the Marne, the French managed to transport to their vulnerable flank near Paris only 11 infantry and 6 cavalry divisions. At present, because of modern rail and motorized transport, we might expect up to one-half of the enemy armed forces opposite the central sector of our western front to shift from one flank to the other within a brief period of time. In addition, air transport in support of operational maneuver essentially reduces to a minimum the time required for new defensive concentrations. Thus, despite the rapid development of enveloping maneuver with motor-mechanized means and aviation, the offensive may nonetheless encounter a continuous front. Besides, it should be taken into account that the means for countering enveloping maneuvers now benefit greatly from obstacles and aviation.

In future war, the "front-against-front" situation should not appear as something unexpected for our operational art, as was the case with the Germans in 1914. It should be recognized as a rather common phenomenon within the dynamic of transforming decisive enveloping maneuvers into equally decisive frontal blows against the entire depths of the enemy's disposition.

This problem brings us to one of the main challenges for contemporary operational art. At issue is the evolution of the second feature of an operation, that is, its distribution in depth. As we have seen, not everything was accomplished in this respect during the maneuver period of the World War (see Part One for the evolution of operational art during the World War). There was, indeed, a chain of interrelated battles, but it was not continuous. Its combat actions did not fill the entire depth of the offensive. In future war, the nature of the operation will evolve in accordance with this very feature of depth. Of course, we have to account for much greater combat densities throughout the operational depths. In fact, even in March 1918, when the German offensive in Picardy penetrated 60 kilometers into the enemy's depths, or at the end of 1918, when combined Entente formations penetrated the German front to a depth of 100 kilometers, continuous battles were waged throughout the offensive depths. Even then combat actions filled the whole depth of the advance.

In future war we will commonly confront such combat depth. It results primarily from the operational deployment in depth of modern combat formations. Combat depth refers not only to the organization of defensive belts, but also to the depth of operational deployments in any situation. The forward line of fighting divisions itself occupies a tactical depth of

6-8 kilometers. Next, we must account for the nearest combat reserves, which constitute a second line 8-10 kilometers behind the first. Farther to the rear, 20-25 kilometers behind the immediate combat reserves, are located additional army-level reserves, which form a third line that might be deployed as separate groups. Finally, all this operational deployment in depth rests on a railroad line located even farther to the rear (25-30 kilometers from the third line, depending on the situation), which can introduce fresh reserves at any time.

Thus, the modern operational deployment of a combat formation can stretch 60-100 kilometers in depth. If this deployment defends, then its depth assumes the form of successive fortified echelons. One must account for the fact that this depth can be continuously supported and constantly reinforced by fresh reserves in case its forward edge is broken or pushed back. The front can be restored by means of reinforcement from the rear or other parts of the fortified front. Reinforcement is now a function of modern permanent mobilization.

It is evident that the entire operational depth must be overcome and traversed with an uninterrupted series of *combat* efforts. Each kilometer must be taken by force.

If combat events during the Marne march-maneuver filled 23 percent of the offensive's time, then at present the same proportion of "combat content" approaches 100 percent. At the beginning of the World War, troops spent more time on the march than in battle. Today, this ratio has changed sharply: troops will spend more time in deployed combat formations than on marches.

These calculations do not exclude the possibility that the enemy might voluntarily cede part of his territory. In this case, the operation would develop by leaps, retaining its combat depth only in certain positions. But such prospects are at present limited. Modern well-developed possibilities for the employment of rear-guard actions, obstacles, chemical means, and aviation necessitate traversing these operational gaps under conditions of great tactical intensity. Moreover, the less territory a country has, the fewer the possibilities there are to yield it.

Thus, as a general tendency, the distribution of an operation in depth will attain full development in future war, just as was the case with the operation's lateral extension during the World War. We can assume that distribution of an operation in depth would be more fully developed in the western European theater of war than in ours. Nevertheless, for us *a future operation will no longer be a broken chain of interrupted battles. It will be a continuous chain of merged combat efforts throughout the entire depths.* It will be a vast sea of fire and combat, spreading across the front as in the World War, but blazing through the entire depths in future war.

Indeed, the history of armed conflict would never have witnessed such combat intensity. The scale itself would constitute a historical milestone, for, once armed combat has encompassed a front and spilled into the depths on land and in the air, there will be no place else to go.

Thus, *depth* is the very essence of the evolving modern operation, and it is this essence that accounts for the operation's enormous intensity.

A modern operation does not constitute a one-act operational effort in a single locale. Modern deep operational deployments require a series of uninterrupted operational efforts that merge into a single whole. In operational terminology, this whole is known as a series of successive operations. However, this understanding is essentially incorrect. *A series of successive operations **is** a modern operation.* Without depth, an operation is deprived of its essence and becomes historically conservative, failing to correspond with the new conditions that define it.

We are confronting the evolutionary shift of the operation into a new dimension, that of depth. It is this dimension that merges a series of successive operational efforts into the general notion of a modern deep operation.*

Under present conditions, we must refer not to a series of successive operations, but to a series *of successive strategic efforts, and to a series of separate campaigns in a single war.* This understanding is historically fundamental to the evolving nature of the operation and its changing forms and methods of conduct. The blunt facts are that *we are facing a new epoch in military art, and that we have to shift from a linear strategy to a deep strategy.*

*Author's Note: Our literature often refers to the future operation as the "spatial operation." This understanding is inexact. Any space on one plane has two dimensions, width and depth. Operations along a front during the World War already reached their maximum lateral spatial limits. Evolving future operations will attain their spatial limits in depth. As this distinguishing feature indicates, the term "deep operation" will best characterize our understanding of the phenomenon.

3. The Contemporary Correlation of Offensive and Defensive Means

The nature of the modern operation confronts any offensive with the necessity to overcome the enormous depth of defensive firepower. This necessity requires first of all material support from all corresponding offensive assets. No matter how mobile and maneuverable the operation on a tactical scale, any formation must finally pierce an opposing front. Tactically, any battle in the end boils down to a frontal attack. It is this attack which determines, completes, and decides everything. Today, the resolution of this primary problem rests on the relative correlation of offensive and defensive means.

Theoretically, the last period of the World War settled this problem in favor of the offensive. It was then that the first indicators of a practical solution appeared. But the World War failed to draw a complete picture of the new offensive means. The exploitation of new technological means for combat (tanks and aviation) did not achieve the intended effect. Their impact failed to exceed tactical application to attain operational results.

Since then, many technological advances have occurred. Modern tanks and combat aircraft are qualitatively advanced weapons, when compared with those of 1918. It is sufficient to refer to the following primary indices which are displayed on the next page in Figure 2.

Moreover, these are not necessarily the most recent indicators, for modern data have a tendency to reflect increases.

Under these conditions, resolution of the competition between defensive and offensive means in favor of the latter becomes even more probable. In respect to quantity, firepower means would naturally be more powerful in the defensive than the offensive. A machine gun and a battery would be always more powerful in the defensive than the offensive. This fact flows not from qualitative differences, but from the nature of targets on the defensive and offensive. On the defensive, batteries and machine guns fire against attacking infantry groups in the open, and they constitute easy targets. In contrast, batteries in the attack operate against dispersed, hidden, and protected field guns and machine guns. These require time, accuracy, and high rates of munitions consumption to suppress. The two situations are absolutely different, and the massing of firepower assets on the offensive remains indispensable.

However qualitatively new technical means of struggle can acquire clear superiority over defensive firepower. In fact, the tank is not a new firepower instrument. It carries the same gun or machine gun which brought about the tank's appearance. The tank is an armored means for their transportation, and this combination adds up to a qualitative solution to the problem of firepower superiority. Mobility, cross-country capability, and armor confer on the machine gun a *new quality* of relative protection from defensive fire, plus the ability to destroy defensive objectives with the sheer weight of armor. The latter possibility constitutes *a new type of blow and attack.* Naturally, a tank-mounted machine gun is more powerful than its dug-in equivalent. And naturally, a tank-mounted field gun has the same superiority over its defensively-emplaced equivalent.

Tank	1918	Current Day
Speed	2–4 km/h	25–40–60 km/h
Range	40–50 km	300 km
Combat Aircraft	1918	Current Day
Horsepower	500–600	3,000
Bomb Load	0.4 T	3–4 T
Speed	120 km/h	400–460 km/h
Range	250–300 km	3,000–4,000 km

Figure 2. Development of Tank and Aircraft Capabilities.
Source: Original to Author.

Fuller's theory is correct to the extent that it argues the tank has changed the correlation between defensive and offensive means in favor of the latter. Moreover, we should bear in mind that mechanization settles another essential question, the avoidance of excessively deep march columns. They are perfect targets for assault aircraft. The technological capabilities inherent in mechanization offer a solution to the tactical problem of vulnerability with a transition to *off-road* tactics. These involve deployed movement in any terrain, a possibility that diminishes the former importance of road networks and alleviates the necessity to move in deep march columns. Off-road movement facilitates maximum rapidity of assault, as well as affording the best passive protection from aerial attack. Off-road tactics are a new feature of modern mechanized formations, and such tactics have great significance for the evolution of contemporary operations. Indeed, off-road tactics alone condition the transition to a new epoch of military art.

All of the above developments increase offensive potential. In fact, qualitative improvements similar to the tank apply to combat aircraft. In the air they carry the same firepower and explosive assets at the disposal of fixed ground defenses. Indeed, the application of these destructive means from high-speed aircraft becomes more powerful than those at the disposal of ground defenses. We should bear in mind that today air-delivered means are more powerful than ground defenses. In this respect, air defense is indeed inferior to air attack.

However, this fact acutely affects both the defense and the offense. Attacking aircraft are equally threatening to the offensive. In this regard the question must be settled by gaining air superiority along the axes of decisive offensive operations. Concentration of massive aviation assets in the air will be as compulsory as concentration of offensive firepower assets on the ground.

In sum, protection from defensive machine gun fire, cross-country mobility, and the capacity to traverse space quickly by air are decisive factors which condition the superiority of new technological offensive means over defensive firepower. The new offensive superiority stems mostly from mobility, which imparts a new quality to firepower in the offense.

The whole evolution of modern military technology flows mostly along the lines of increasing and perfecting this mobility. Everything that increases mobility enriches offensive potential. Defensive potential can be increased only by improving firepower. But, with reference to increasing rates of fire, everything was already achieved in the World War, when machine guns came into use with the infantry. The only unresolved problem is the automation of artillery. Once rapid-fire antitank and antiaircraft guns are introduced, other means must be sought to counter the offensive. In general, defensive potential has reached its full peak.

It is necessary to note that the defense will enlist modern science and technology to counter the offense in various ways, including engineering assets, chemicals, obstacles, mine fields, and even electricity and radio for long-range disruption and destruction. However, only a stabilized continuous front can support the widespread use of these modern technological means. Meanwhile, the development of modern high speed combat means, including aviation and motor-mechanized assets, can to a significant degree condition the mobility of military actions.

Evolving science and technology do hold prospects for countering the offensive. Still, it is evident today that the offensive leads in the development of technological combat means, while the development of defensive means occurs only in response.

That this offensive superiority has affected the European general staffs is evident from the appearance of modern permanent fortification systems. The eastern border of France is now a continuous line of concrete fortifications, with electrified fields of death guarding the approaches. The Germans now build similar fortifications in the remilitarized Rhineland. Overcoming a belt of concrete fortifications is of course impossible for modern offensive means. If the art of fortification evolves to fast-hardening concrete, making it possible to build concrete fortifications quickly during the course of maneuver, then the probability increases that military art will confront the new problem of scientifically and technologically advanced trench warfare. It is difficult to predict the evolving nature of such a confrontation, but it is possible to assume that its prerequisites are rooted in the possibility for a second imperialist war on the western European sub-continent. Under the insurmountable conditions of new-style trench warfare, such a war would be doomed to failure and of course would promote the development of the conflict into a civil war on a global scale.

There are no prerequisites for such a positional front in our eastern European theater of military actions. However, such a front with its qualitatively different character could arise in isolated sectors. Therefore, it is necessary to realize the prospects of storming a concrete line front from the air and not from the ground.

Airborne forces must play an important role in the future. It would be hard to overestimate their significance in the evolution of operational art.

Under modern conditions of colossal technological progress and corresponding prospects for our future development, we should never be short-sighted and lag behind. The competition between offensive and defensive means affords a vast field for research and experimentation. It is necessary to keep in mind that combat means should always be viewed with respect to the means for countering them. In an evaluation of combat means, one can never assert that their inherent characteristics preclude the possibility of their being overcome by countermeasures. Countermeasures will never remain on a level at which they become less suited for further development than offensive means. The adversary who must defend himself will naturally employ all possible means of resistance. The course of conflict can present many new possibilities that make insufficient a blow by modern offensive means.

From this perspective, we must therefore recognize that possible countermeasures will make the application of force by modern offensive means less certain and convincing than the benefits of range and speed. This situation might require the further innovation of offensive means. Far-sighted and progressive technical thinking must keep this fact in mind.

One thing, however, is apparent: the present tendency favoring the superiority of offensive over defensive means is growing more palpable. Under those political conditions which determine the nature of our future war, this circumstance affords a material foundation for the possibility of overcoming a firepower-intensive front and for producing a decisive outcome with deep offensive operations.

4. The Organization of the Offensive in Depth

Combat means are a necessary material prerequisite for solving problems, but they cannot solve problems on their own. There are many instances in the history of military art when new combat means failed to produce the desired effect. They were employed in outmoded combat formations in accordance with dated methods. Such was the case, for example, when rifled field guns were left at the rear of march columns.

New armament requires new forms of combat employment. Tactics settled this question through transition to combat groupings and deep battle. However, the control of large troop formations has lagged behind, mired in an earlier stage of historical evolution. Once the objective of an offensive displays great defending depth, the operational deployment of an offensive attack formation requires essential changes. A single line of deployed armies would hardly be able to solve the *new problem of the deep offensive.* One can definitely assert that linear strategy's *single wave of operational efforts will not solve anything.* It would powerlessly dash itself against the depths of modern defenses.

This problem brings us to the central question of fashioning a deep strategy for the present epoch. It is necessary to perceive the character of the modern defensive depth: resistance tends to increase and to attain its culminating point or strategic zenith when the attacker is close to his aim and the defender must put everything on the table to save his position. Because the belligerents' tenets are incompatible, and because there is no reconciliation in a conflict for political and economic independence, resistance can display enormous strength at the last stage of an operation.

Even during the World War, when the contradictions within imperialism were acute, operations in 1914 developed along a curve of rising combat efforts. This fact escaped the Germans, who entered the first frontier battles with a high operational intensity, but who approached the Marne poorly prepared to confront increased Anglo-French resistance.

Nor did we take this rising resistance curve into account during our offensive of 1920 to the Vistula. After forcing the Nieman, it was even planned to reduce the strength of the western front armies, since completion of the campaign seemed assured at the initial stage of the offensive. The operational forecast did not envision a battle of enormous intensity on the Vistula, and this was a bitter miscalculation that testified to a deep misunderstanding of contemporary operational dynamics.

Offensive exhaustion finds its causes less in the self-induced expenditure of attacking power than it does in growing defensive resistance. An obvious example would be a situation in which linear strategy's offensive front simply repelled the enemy rather than capturing or destroying him. Thus, the failure of this strategy to bring about the destruction of the enemy's vital force would permit the same retreating enemy to occupy an operationally advantageous position. Consequently, at the very culminating point of the operation, the defender would be much stronger than at the initiation of hostilities. Meanwhile, the attackers would carelessly approach this strategic Rubicon, assuming that the final moment of the operation would be the easiest. Such would be a fatal mistake. It is always the first step that is easier, because it is assured by advance planning and the preliminary grouping of forces.

Difficulties must be expected during the course of an operation, since all details cannot be foreseen. *One should expect the greatest tension and crisis at the final stage of an operation.* The essence of the art of operational leadership lies in the ability to approach this decisive moment in full awareness of the situation, with a fresh wave of operational efforts, and forearmed with all the necessary forces and means to put a crushing end to the operation.

The leader is doomed who would presently try to approach the Marne or Vistula as in 1914 and 1920. His end would be inglorious, no matter how grand along the way were his offensive operational achievements. Moreover, the grander these achievements might be, the graver would be the catastrophe if forecasting were not applied to the final stage of the operation.

A modern operation is an operation in depth. It must be planned for the entire depth, and it must be prepared to overcome the entire depth. Moreover, it must be anticipated that the intensity of resistance within this depth tends to increase and grow denser from front to rear.

In elaborating the deep offensive operation, contemporary operational art encounters the novel problem of structuring offensive formations. One thing is clear: linear strategy with its *single wave of operational efforts is incapable of dealing with this offensive problem.* The solution is to be found in accordance with the new ways of evolving operational art. Meanwhile, one proposition of traditional military theory must be discarded along the way. Before anything else, we must abandon the proposition that *strategy achieves its aims in accordance with the principle of simultaneity of actions.* This proposition, which enjoys popularity even now, dates to Napoleon's age. It lost relevance some time ago under modern conditions. Clausewitz referred to it several times:

> "In tactics, when forces are gradually introduced into battle, main decisions are postponed until the end, whereas in strategy the law of simultaneous engagement of all forces almost always strives for decision at the beginning of a larger action...."

> "Tactics allow gradual introduction of forces into battle, while strategy makes its demands immediately and simultaneously...."

> "Strategically one must engage the largest number of possible forces, their engagement must be simultaneous...."

> "Strategy cannot recognize time as its ally, and for this or that aim introduce forces into an affair gradually and incrementally. All available forces assigned to achieve a strategic aim must be engaged simultaneously...."

"In strategy dispersed efforts contradict the essence of the aim; all available forces must be engaged simultaneously."

This theory was correct for Napoleon's age, as well as for the outset of linear strategy in Moltke's era, when an operation still generally led to a one-act main battle that was decided by a single wave of operational efforts. However, this theory did not correspond with the new conditions of armed conflict during the epoch of imperialism. Its death throes were already perceptible during the last decades of the nineteenth century.

During the second half of the Franco-Prussian War of 1870-1871, after the fall of the Second Empire, the Prussians had insufficient forces to engage in a new struggle against a reorganized French army. But the treacherous counterrevolutionary French bourgeoisie lent Moltke a helping hand by making peace with Bismarck over the head of the French National Guard. It is now difficult to speculate how a renewed war between France and Prussia would have ended under different circumstances. But Engels described its possible outcome in the following way: "The French position was very strong despite their recent defeats. If we could be sure that Paris might have held out until late February [1871], we would be inclined to speculate that France might have emerged as the victor..."(Engels, *Stati o voine,* Izd. 1924 g., 182, 195.)

Even then there were the first signs of permanent mobilization and of the impossibility of achieving strategic decision by sheer simultaneity of a single effort. Moltke realized he was facing a new phenomenon in the history of armed conflict. Later he said several times: "This war (i.e., a continuation of the 1870-71 war after Sedan) astonished us so much that the question it posited should be studied many years." Indeed, the question was worth study. The appearance of new armed forces after a first-line enemy army had ceased to exist indicated that strategy might not achieve its future aims with one first-line army deployed at the beginning of a war. The introduction from the depths of a second and possibly even a third-line army might be necessary. In Moltke's vague premonition there was a convincing hint of the epoch of deep strategy.

In his famous speech of 1890 to the German Reichstag, Moltke said: "If a war, which for more than ten years has been hanging over our heads like the sword of Damocles, finally breaks out, no one can predict its duration and outcome. The greatest European states, armed as never before, would enter the war against each other. *Not one of them would be crushed during one or two campaigns, so that it would recognize itself as defeated, so that it would be forced to conclude a harsh peace, so that it would not reaffirm its strength and resurrect the fight."* (emphasis added, G. I.)

This was a different Moltke, a strategist of the new epoch. But new views were incapable of disproving old theory. Historical experience went unnoticed. Even at the beginning of the twentieth century, Foch wrote the following in his *Principles of War:* "Within strategy the law of coinciding efforts governs, not the tactical law for the gradual reinforcement of effort." This view was already incorrect during the war of 1870-1871, and all the more so during the war of 1914-1918. At present this proposition is absolutely incompatible with the new character of the deep offensive operation.

In this regard, that, which during the second period of the Franco-Prussian War of 1870-71 was perceived only on a strategic scale, manifested itself operationally during the World War and later in the contemporary field of operational art. *A modern multi-act deep operation cannot be decided by a single simultaneous blow of coinciding efforts. It requires deep operational reinforcement of these efforts, which expand in proximity to the highest point for attainment of victory.*

Deeply-echeloned resistance causes equally deep offensive echelonment. The offensive should resemble a series of waves striking a coastline with growing intensity, trying to ruin it and wash it away with continuous blows from the depths.

A modern operation essentially elicits distributed efforts in time, thereby conditioning strategy. This observation was proved by events during both the World War and our civil war. *But of course it would be wrong to understand* that the Germans in the frontier battles of 1914 and we in the battle on the Auta River in 1920 engaged too many forces at once, and that these forces ought to have been engaged gradually. All available forces should be engaged during initial operations in accordance with the correlation of belligerent forces. But the essence of the question is the necessity beforehand to organize the deep echelonment of additional efforts. At the decisive moment of the operation, the object is that additional forces and means arrive in the appropriate groupings to facilitate final attainment of victory.

Modern operational echelonment of efforts in depth does not mean engagement of these efforts either piecemeal or in operational packets. *Modern operational echelonment is the sequential and continuous increase of operational efforts aimed at breaking enemy resistance through its whole depth.* The greater the resistance in depth, and the greater its intensity, then the greater must be the echelonment of the offensive's operational depth.

While deploying for a modern deep operation, it is necessary to calculate forces and means both along the linear dimension of a front and in the new dimension of depth.

The problem of deep operational offensive deployment challenges another time-worn proposition, the idea of so-called strategic reserves. As long as strategy solved a problem with a single simultaneous effort, no reserves were needed. Clausewitz described the idea of strategic reserves as senseless, calling them unnecessary, useless, and even harmful. He insisted that all strategic efforts be compressed into one action during one moment. He wrote: "The idea of holding back prepared forces for use after attainment of the general aim is impossible to recognize as anything but absurd."

As long as the general aim was achieved by a single act in Napoleon's era, this proposition was correct. However, doubt set in during the second half of the Franco-Prussian War of 1870-71. By the beginning of the twentieth century, the proposition simply became incorrect.

To a certain extent, Schlieffen had foreseen this problem. He insisted on having a strong reserve army behind the German right wing during the advance on Paris. But his motives were different. He needed operational reserves during the offensive to extend his right flank in case additional forces were required to complete envelopment of the enemy. In the end, Schlieffen's reserve would enter the same line as the advancing front.

Under modern conditions, operational reserves are required not to extend flanks, although such action might still be necessary at the beginning of a war. In general, flanks have already reached the limits of their lateral extension, so reserves are now necessary for reinforcement of operational efforts aimed at breaking the entire depth of enemy resistance. Now, the very notion of operational and strategic reserves involves the development of operational echelons. As armed conflict evolves to the future, the silhouettes of analogous strategic echelons will appear behind these operational echelons. Of course, this development would lead to further increases in the strength of armed forces, thereby disproving any theory about small professional armies as conservative and nonsensical.

The growing strength of armies during the epoch of imperialism answered the requirement of linear strategy for the broadest possible enveloping offensive front. Now, the growing strength of armies is a function of deep strategy, which requires strong operational echelons in depth and deployment of the offensive in depth. These developments testify to the grand scope of contemporary armed conflict. They also disclose the whole evolving character of the operation during the emerging epoch of deep strategy.

5. The Entry in Depth into the Contemporary Operation

New requirements give rise to new historical phenomena, but these phenomena are also predetermined by a number of new prerequisites. In 1866, when for the first time Moltke deployed Prussian armies across a 400-kilometer front, this operational phenomenon corresponded with the new character of armed conflict. But this phenomenon was also predetermined by new objective conditions, including railroads. Still, as Schlichting has noted, Moltke harbored strong misgivings about such broadly-based deployments.

So it is today, when *deep deployment* generates apprehension and even fear. But, whether we like it or not, such deployment is inevitable. At present, a number of objective prerequisites predetermine deep deployment. It flows from the nature of future war, which will generate a conflict of immense intensity. No country entering this conflict will limit mobilization capacity to the first echelon of a mobilized cadre-based regular army. Further, no country at the outset of war will have the capacity of simultaneously concentrating for immediate combat action all forces capable of mobilization. To do so would require the postponement of hostilities and the withdrawal of one's own deployments deep into the country's interior to protect them from piecemeal destruction. In this case, a weaker enemy with fewer forces to deploy would paradoxically be stronger at the very beginning of conflict. However, there are few who would dare test this proposition. It is very evident that *sequential permanent mobilization* leads to a sequential buildup of efforts.

First-line forces would be followed by second- and third-line forces, a situation that predetermines ground force entry into war by deep strategic echelons. This inevitable scheme for entry by depth into a future war is reflected by an army's contemporary peacetime deployments. How else can we explain the existence of a special French covering army *(l'armee couverture)* on the Rhine? In fact, this army constitutes the French first strategic echelon, behind which the main mass of the armed forces will deploy for combat in second and subsequent echelons. German occupation of the Rhineland definitely aims at the concentration there of the same kind of covering army, but one suited to become the first operational echelon of an invading army.

The deeper and vaster a country's territory, the greater is its mobilization potential, the more powerful is its capacity for combat intensity, and the broader is its scope for deeply-echeloned strategic efforts. These conditions apply to our country. They amount to a powerful advantage that facilitates *the maximum increase of efforts at the last decisive moment of conflict.* In comparison, the Baltic countries are much smaller in territory and weaker in mobilization potential. Their mobilization intensity at the beginning of any conflict would be close to its peak. The all-important gradual buildup of efforts would occur in the Baltics on a much reduced scale, unless large imperialist countries seriously rendered them assistance with forces and means.

The echeloned entry of armed forces into a war is a function of both strategic and operational necessity. Prerequisites for this necessity flow from material factors in evolving contemporary combat technologies. The essence of the technological evolution of modern armaments lies in the impulse for greater range and range of action for effect. Everything boils down to inflicting destruction at the greatest possible range. The entire significance of combat aviation lies in the capability to cover distances quickly. The same holds true for motor-mechanized means.

The evolution of firearms followed the same path. It is worth noting that during the second half of the nineteenth century the development of firearms focused on range and rates of fire. Before the World War, at the turn of the twentieth century, the focus shifted mostly to improved rates of fire, while ranges remained at previous levels. After having attained maximum rates of fire with the machine gun, technological evolution during and after the World War emphasized increased ranges. Machine guns were fitted with inclinometers so they might fire at distant targets from concealed positions. Improvements in field artillery increased its range to 12-20 kilometers. All these developments held decisive significance for evolving tactical forms of battle.

Historical evolution demonstrates that the increased range of weapons during Moltke's era accounted for transition from Napoleonic-style concentration of all forces before battle to the meeting engagement from the march. Now, because of still greater ranges, we face further evolution of the meeting engagement. During the second half of the nineteenth century, when the range of fire came to equal the range of vision, engagement from the march was the direct result. Now, however, the range of fire is much greater than the range of vision in the field. This development means *that modern battle will commence at great distances.* It also means that present tactical march security, positioned 5-6 kilometers in advance of main forces on the move, in fact secures nothing either from remote firepower assets or from the sudden onslaught of motor-mechanized troops. This assertion does not even account for attack aviation, which can transit immense distances in another dimension. March security forces are no longer capable of fulfilling their role as the advance guard to cover the deployment of main forces for battle. March security now constitutes local security only. In addition, the increased depth of movement columns requires more time and space for main force deployments in appropriate groupings for battle. At the turn of the twentieth century, [General Hippolyte] Langlois, who was developing a theory for evolving artillery employment, wrote: "We must press our advance guard forward not several kilometers, but several miles, up to a distance of 1-1.5 march traverses."

Modern combat ranges have increased markedly. They require the forward deployment of movement security at distances of at least 20-30 kilometers, which is the depth required for the deployment of a contemporary reinforced division. This requirement essentially demands the widespread use of a system of forward reconnaissance detachments. Such a system would screen the advance guard, which itself moves 5-6 kilometers ahead of the main force, and assume the functions of reconnaissance and security. Without these functions, the advance guard simply becomes the first echelon within the march column. The new reconnaissance and security detachments must be sufficiently strong to perform their functions. However, such changes would settle the security issue only on a tactical scale, on the level of security precautions along a given route of advance.

The contemporary army commander who actually desires to control a modern deep operation must first of all provide for the timely deployment of his forces and their entry into battle in a grouping that accords with his intent. He needs an instrument for operational security that consists of powerful mobile formations, chiefly motor-mechanized units and cavalry, pushed forward one-two and even more traverses.

In contemporary circumstances we return to the Napoleonic phenomenon of the *army advance guard* as the first echelon in the march, but with a completely different qualitative significance during the emerging epoch of deep strategy. This dialectical transformation closes the evolutionary circle for offensive operational deployments. The essence of this transformation means that the notion of a meeting engagement has reached its zenith, moving from the field of tactics into the operational sphere. As a rule, a meeting engagement is tactically possible only for forward units of the advance guard echelon. But operationally, the engagement becomes a *meeting battle* when the advance guard echelon functions as an army-level advance guard. This shift means that the contemporary operational formation for the offensive must inevitably be deeply echeloned.

It is possible to approach this new phenomenon with apprehension and misgivings, even though it is rooted in new requirements for the contemporary operation. Deep echelonment is inevitable, for it has been predetermined by a number of objective conditions.

It must be taken into account that modern combat means are very diverse, with reference to their speed, range, and effects. Aviation naturally occupies first place in range and the ability to cover long distances. Earthbound enemies will not even have commenced firing when this service arm begins attacking during the first hours of war at very long range. Powerful and massed combat aviation will naturally be the first factor with combat impact.

Aviation will be immediately followed on the ground by everything that is mobile and easy to displace forward, especially motor-mechanized units and modern mechanized cavalry. While the core of the first-line army laboriously completes its complex mobilization, the mission of these forward horse and motorized units will be to disrupt enemy concentration and then occupy an advantageous jumping off position for transition to the general offensive. These mobile units will constitute the first ground advance guard echelon.

Finally, the main body of combined-arms infantry formations will enter the theater of military actions. But this mass of troops will not be able to form one line immediately. Because modern railroads have grown more slowly than the armed forces, the railroads will not be able to transit all troops immediately and completely. The result will be a prolonged period for the concentration of all forces in the theater of war. When the majority has completed transit, it will begin operations as soon as possible. Those forces arriving subsequently will begin operations later. Thus, the main body of forces will deploy in two phases to comprise the second and the third operational echelons.

When this entire in-depth system of the first strategic echelon begins to move, the outline of a second strategic echelon will take shape in the strategic depths of the country. This echelon will be comprised of mobilizing second-line troops.

If all of the above does not signify the onset of an epoch of deep strategy, then one has to doubt the very notion of depth.

The physical boundaries for entry by depth into an operation will stretch to immense distances, (see Figure 3). Aviation will immediately operate at its maximum range. Motor-mechanized units and cavalry will rapidly advance 2-4 traverses forward (about 100 kilometers). The attacking first echelon forces of the main body of troops will occupy a depth of 75 kilometers, provided each division has its own road (which cannot be always ensured). Finally, second echelon forces of the main body of troops will be one traverse behind the first echelon. The second echelon will extend across a wider front than the first, and it will occupy a depth of 50 kilometers.

In general, the entire first strategic echelon would occupy an immense depth of 250-300 kilometers on the ground. However, such depth cannot be ensured by modern conditions of deployment.

Predicting twentieth-century conditions, Schlichting wrote: "the strategic deployment of an army will be only several short traverses removed from the first decisive main battle." Meanwhile, [Jules-Louis] Lewal predicted that "in future war contact would occur spontaneously right at railroad station debarkation points." Under present conditions, when troops in heightened mobilization readiness are located close to the border, and when covering forces are concentrated closer to the border, military operations will practically start right on the spot. Long 300-kilometer marches through the depths will be unnecessary.

The above-mentioned deployments are perfectly obvious on the Franco-German border. General Debeney has said: "At the beginning of a future war France and Germany will already be in contact, since French garrisons are deployed not more than 20 kilometers from German border-guards entrenched in the woods. The battle field will not afford sufficient space to permit motorized troops to use their speed." In addition, shallow depths will not permit a number of small states to develop deployments in depth. In such situations, the operational offensive depth will not reach its full potential in space. Operational echelons will enter the operation from one line.

Regardless of circumstance, the last echelons will be peacefully marching in the deep rear, perceiving during their advance a threat only from the air and the intensity of supply and evacuation activities, while the first operational echelons will already be engaged in fierce battles, during which much will be resolved. It will be difficult to predict not only when and where this grand operation will take place, but also when and where to draw any noticeable boundary between the operation and main battle. We will be crawling into this battle, when in essence the first bomb dropped in the deep rear or the first shot fired will already have signaled the initiation of this grand operation.

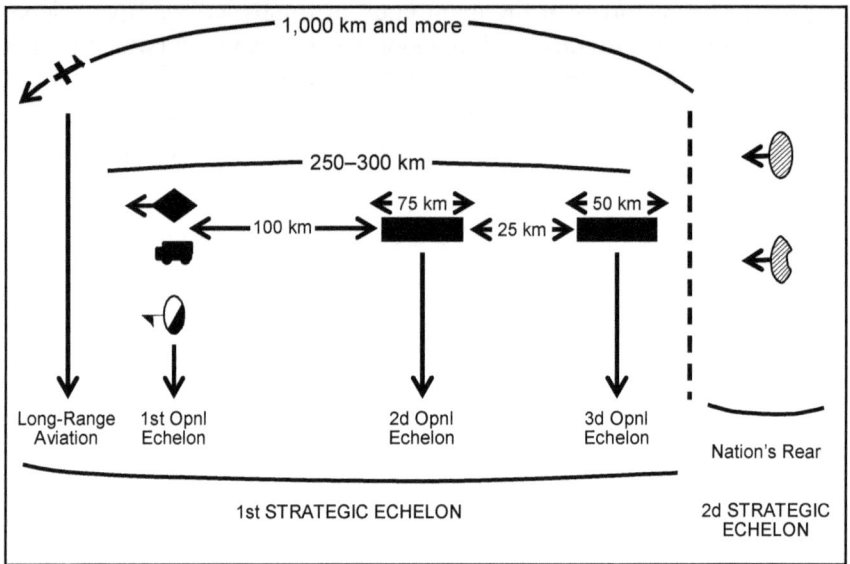

Figure 3. Entry in Depth into a Modern Operation.
Source: Original to Author.

During the epoch of linear strategy, main battle emanated organically from an operation, whereas *during the epoch of deep strategy the operation and main battle will organically merge.* Any boundaries in time and space will disappear. During a single expanding torrent of operational efforts, modern main battle will envelop a front and find its conclusion in the depths. Thus, wave after wave will break against the approaching enemy front, which will obviously be similarly deployed. From this situation arises the conclusion that *final success will reside with the side having the deeper operational deployments.*

The moment is inevitable, when all these waves will co-mingle in a single squall of fronts directly confronting each other. At this point, perhaps the development of the operation will once again produce a linear front and linear strategy. But, also at this stage, which could come naturally and soon under modern conditions, the evolution of operational art might require a different resolution, with a deep frontal blow from the depths into the depths. Here the requirement for deep offensive deployments would become even more acute. The result would be a new operational solution for the problem of conducting a breakthrough during the emerging epoch of deep strategy.

6. The In-Depth Breakthrough and Destruction of the Front

During the epoch of linear strategy, operational art reached its own self-negation when front confronted front, thus necessitating a breakthrough. The problem could not be resolved operationally on the basis of linear strategy. This quandary elicited the appearance of new technological means. It also raised to a new level the technique for the tactical organization of the offensive and created preconditions for tactical resolution of the problem. But linear strategy all the same could not resolve the operational problem of breaching and destroying a front. So, operational art had to look for new methods, it had to step forward into a new epoch. But an imperialist war of attrition and exhaustion did not provide the appropriate conditions.

The new nature of future war with its decisive destructive operations has advanced a new kind of resolution for the central problem of contemporary military art. *A front must be broken by means of a decisive operation. A front must be broken and totally crushed throughout its entire depth.* Deep strategy will pass the test of historical maturity. If this strategy has been predetermined by many contemporary objective conditions, at the same time it has been evoked by requirements for decisively and fully overcoming the frontal phenomenon.

New forms of deep battle are conditioned by the widespread tactical employment of modern technological means for combat (tanks, long-range artillery, and short-range aviation). These means can solve the breakthrough problem on a tactical scale. But they can only breach the tactical depths of modern defenses. Tactical means remain unable to produce operational decision, although they lead to it.

Deep tactical efforts must still evolve into a deep operational breakthrough. Operational art during the epoch of deep strategy must resolve this basic problem. All the attainments of deep tactics will become superfluous if this problem is not resolved on an operational scale. One must understand that the first attack echelon for breaching a front is capable of fulfilling its mission only on a tactical scale. No matter how grand the success, the first echelon by itself cannot transform tactical results into operational results by rushing through the broken door to crush enemy resistance through the entire operational depths. The first attack echelon cannot resolve this problem, for strong springs offer resistance inside the broken door, and it has to be held against slamming shut. This combat mission remains the duty of the first attack echelon. But, if no one takes advantage of the tactical breach made by the first echelon, *if no one comes from the operational depths to prolong the depth-to-depth blow,* and if tactical success doesn't become operational, the breach will soon close. All the tactical efforts of the first attack echelon will have been wasted. After the attackers had exhausted themselves, nothing would remain, except a belly-like protrusion in the offensive front. Such would be a continuation of the system of senseless and exhausting frontal attacks of self-attrition to which linear strategy gave birth in 1918.

The modern breakthrough can and must be undertaken not only when there are sufficient forces and means to pierce a front, but also when there are sufficient forces to extend the rupture in depth for destruction of enemy resistance throughout the entire depths. Undertaking a breakthrough operation is wasted effort unless there is sufficient strength for its development. It is senseless to break down a door if there is no one to go through it.

A modern deep breakthrough essentially requires two operational assault echelons: an attack echelon for breaching a front tactically; and a breakthrough echelon for inflicting a depth-to-depth blow to shatter and crush enemy resistance through the entire operational depth, (see Figure 4). Both echelons retain their own internal tactical echelonment. This deployment in depth for a breakthrough operation resolves the main problem of modern operational art, i.e. the problem of a decisive, full, and deep breakthrough to bring about the front's complete destruction. Depth of formation remains essential not only for breaching fortified defensive belts, but also for launching any frontal blow that arises during the course of frontal main battle. In contemporary operational perspective, the only side that can count on final success is the side with the deeper formation, and the side with the more powerful echelons.

At the turn of the twentieth century, during the golden age of linear strategy, Schlieffen taught that victory belonged to the side with the longer and stronger flank. Now, we must refute this teaching in modern operational perspective with the proposition that under the contemporary conditions of deep strategy *victory belongs to the side with the deeper front and the more powerful deep echelons.* In a relative sense, we must keep in mind the obvious prospect for larger contemporary armed forces, while discarding as absurd various theories about small professional armies.

It is now necessary only to depict the entire in-depth scheme for a modern breakthrough operation. The operational art born of deep strategy will come into its own when waves of operational effort from the depths combine with a first advance guard echelon already engaged in main battle to produce a general squall and when, in consequence, two fronts confront each other without possibility for envelopment. The fast-moving *advance guard echelon* of motor-mechanized units and cavalry must be withdrawn early from the combat front because their long-range effects are no longer suited to the situation. There will be insufficient maneuver space, and they will have fulfilled their mission as an army-level advance guard. These units will now move to the flank on the way to redeployment in the rear of the offensive operational formation.

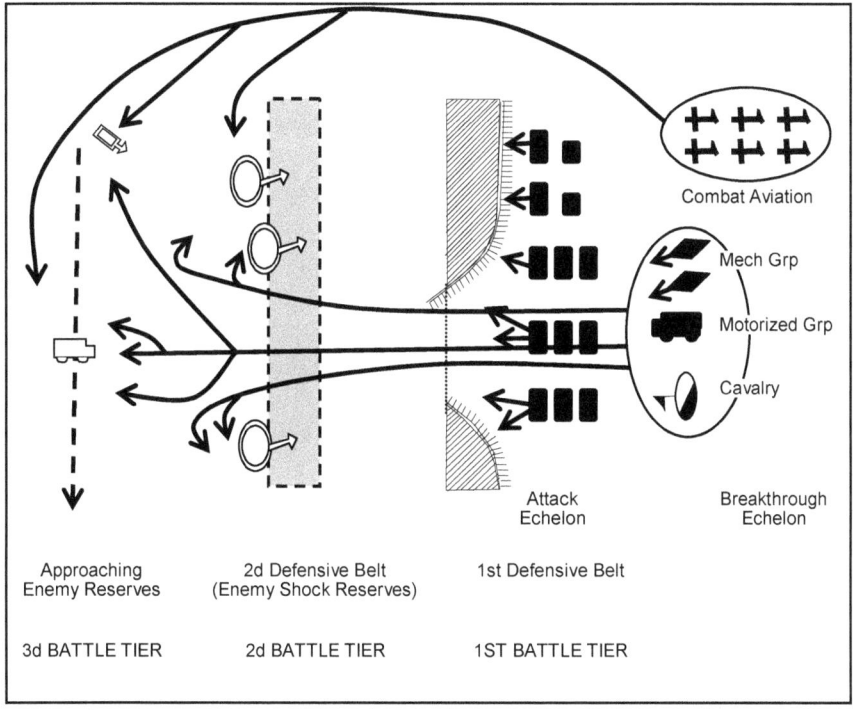

Figure 4. The Deep Operation for Penetrating and Crushing a Front.
Source: Original to Author.

They will be replaced by advancing echelons of combined arms infantry formations, the effects of which are more appropriate to combat against a front. These formations comprise the *attack echelon,* since they constitute a tightly-deployed operational phalanx, armed with numerous tanks, highly-effective heavy artillery, and short-range combat aviation. They will be followed by a *breakthrough echelon* of fast moving units tailored in advance as an offensive operational formation. It would consist of large independent motorized, mechanized, and cavalry formations supported by large masses of long-range combat aviation. The units in the lead at the beginning of the operation would now fall to last in the operational formation, whereas the ones which were last in the approach march would now become first in the attack.

This is the operational formation for the beginning of a deep breakthrough operation. It will display deep operational offensive deployments aimed at prolonging and developing depth-to-depth blows. This formation has nothing in common with the echeloned offensive breakthroughs of 1918. During the March 1918 offensive, the Eighteenth German Army had 12 divisions in the first echelon, 8 divisions in the second, and 4 divisions in the third. During the May 1918 offensive, the Seventh German Army had 14 divisions in the first echelon, 5 divisions in the second, and 6 divisions in the third. During these offensives, each advancing division had only 3 kilometers of depth, while succeeding echelons had to replace and supply forward fighting units while pressing the offensive forward along a common frontline. The piling up of these echelons was reminiscent of the strategy of a stampeding buffalo herd which could not understand the requirements for an actual frontal breakthrough. That is, for tactical efforts to become operational, the blows must be prolonged and developed from the depths into the depths. In 1918, when there were no independent motor-mechanized units, and when cavalry had practically ceased to exist, resolution of the situation could not be assured. The breakthrough echelon must be faster than the attacking echelon in order to overtake and pass through it. Therefore, the breakthrough echelon could not be comprised of infantry. The breakthroughs of 1918 were tactical phenomena that could not be transformed into an operation. They were unable to posit the aims appropriate to the operational art of a deep strategy.

A contemporary deep breakthrough operation pursues the aim of simultaneously breaching and crushing the entire operational depths of the resistance. But operational simultaneity cannot be equated with tactical simultaneity. There is a difference in timing for effect. This difference is determined tactically by breaching the depth of the first defensive belt. After the attack echelon fulfills its tactical mission by breaching the enemy front, the breakthrough echelon pours through the breach from the operational depths. In the air, long-range combat aviation will outpace ground forces to preclude entry of enemy reserves into the breached sector. At the same time, airborne units will land in the enemy rear to become the first messengers of death. Simultaneously on land, a huge multi-wave, lava-like mass of fast-moving tanks, self-propelled artillery, and infantry in armored transporters will rush through the tactical breach in the front. These forces will destroy the last bottlenecks within the open breach. They will be followed by modern cavalry, "the arm of glory," preserved by history. Finally, after roads are restored, numerous columns of motorized forces will enter action. Each component part of the breakthrough echelon will have its own role to play in the open breach. The breakthrough will occur simultaneously in several sectors of the front.

All these factors will prolong and develop the depth-to-depth blow. The larger the breakthrough echelon, the greater will be the depth of its objectives. In all instances, the offensive blow must traverse the entire depth of enemy resistance to fulfill the operational breakthrough mission. While the attack echelon continues to wage fierce battle in the breakthrough sector, on another level, perhaps even at prescribed tiers within the defensive depths, the breakthrough echelon will begin actions for encirclement and destruction. In operational perspective, these actions would become a *new grand multi-level battle waged on several tiers within the operational depths.*

This battle will resurrect "Cannae" on the new basis of deep strategy. In fact, an entire "Cannae" system would appear, with some battles under way, others on the verge of beginning, and still others completed. The operational breakthrough of a front will be decided by the decisive shattering and destruction of resistance. Never has a strategy for annihilation enjoyed such splendid prerequisites for its full realization. This projection solves one of the grander problems in the evolving nature of modern operational maneuver.

The practice of armed combat and the theory of military art have thus far distinguished between two main types of operational maneuver. The first, characteristic of Napoleon's era, was maneuver along interior lines for a concentrated blow against a single position. The second, characteristic of the era of linear strategy, was maneuver along exterior lines for an enveloping blow from various directions. These two types of maneuver were contrasted with each other, and to a certain extent were considered operational antipodes.

Clausewitz characterized them as follows: "In strategic maneuver two opposites are encountered, and they seem to be completely separate types of maneuver. The first opposite is action along either interior or exterior lines. The second is concentration of forces either at one point or along many points." But historical evolution gives rise to the new by combining and transforming varied things.

A contemporary operation for a deep breakthrough is a unique combination of two types of maneuver. The attack echelon, which breaks the front, occupies a broad continuous line and operates along exterior operational lines. The breakthrough echelon operates on interior operational lines to inflict a concentrated depth-to-depth blow. *Thus, the epoch of deep strategy leads to a synthesis of two types of maneuver, or of two historical schools of military art.*

So, we discard the frequently voiced and non-dialectical idea that maneuvers of envelopment and encirclement have ceased to exist. Such opinions find no reflection in the foundations for the evolving nature of a contemporary operation. These opinions fail to see an operation in its two dimensions, i.e. along a front and in depth; they remain conservatively wedded to linear strategy.

A frontal blow is naturally the main form of action for the first attack echelon. But in itself, the frontal blow resolves nothing unless the attack echelon's tactical efforts become operational. But this transformation can be achieved only by inflicting a blow along interior lines from depth to depth, in order to envelop, encircle, and destroy the enemy.

Of course such maneuver does not occur along the linear front, but is transferred with great intensity into the combat front's depths. Here maneuver is fully reborn in great scope with new content. Here maneuver promises a golden age of deep strategy as the art of splendid maneuvers and crushing blows in depth.

Thus, the epoch of deep strategy will complete the evolution of military art.

7. The Art of Commanding a Deep Operation

Naturally the new character of the deep operation requires a new way of conducting it. As the art of conducting operations, operational art faces a number of new problems. During the emerging epoch, when the operation flows organically from main battle, when these two notions become a unified whole within the phenomenon of deep strategy, and when they are not bounded by space and time, we have to abandon the proposition that, "the moment the enemy approaches closely enough to offer general decisive battle, the time for strategy is over, and it can take a rest." (Clausewitz)

If operational art excluded main battle from its sphere of competence, operational art would become self-destructive and useless. Such a situation occurred in 1914 during the march to the Marne and in 1920 during our march to the Vistula. In own his time, Schlieffen wrote, "*Long before a probable clash with the enemy,* the most important task of a leader can be considered fulfilled if he assigns roads, ways, and directions of movement for all his armies and corps." (emphasis added) On the basis of such advice, we have seen the degree to which uninformed straight-line offensive operations degenerated in 1914 and 1920. But the inertia of conservative theory is great. The experience of the World War has not been fully studied. Even now the French Lieutenant Colonel Duffour writes, "The essence of strategic maneuver lies in the formation of columns with various tasks and their direction to the general objective. From the moment a maneuver moves from intent to execution, it is expressed only as a designated direction and allocation of forces for units on an assigned axis. The execution of maneuver is the selection of directions and the distribution of forces and means among columns." (Diuffur, *Lektsii, chitannye vo frantsuzskoi vysshei voennoi shkole.*)

Thus, in Duffour's opinion, strategy (or operational art, to be more exact) deals with forming and directing columns and disappears as soon as the columns engage in battle. The operation would evidently unfold along rigidly assigned directions until it became a simple wall for pressing back the retreating enemy instead of destroying him as required by the essence of combat. This method brought failure to the large offensive operations of 1914 and 1920. Under emerging conditions, the same method will lead to anarchy. The situation is reminiscent of a clown wearing large shoes, who wants to grab hold of a ball on the ground. But, to the amusement of the crowd, each time he approaches the ball, his oversize shoes inadvertently kick the ball farther away. In the same way, outmoded military theory now attempts to amuse history, for transference of outdated ideas to a new epoch with absolutely new conditions constitutes a historical joke. Armed conflict waged for grand purposes imposes greater demands on itself.

During the epoch of deep strategy, a deep multi-act, multi-level main battle incorporating all an operation's phenomena will lie from beginning to end within modern operational art's sphere of competence. Otherwise there absolutely cannot be any operational art.

The formation and dispatch of columns will scarcely be its most significant aspect. In fashioning a deep operation, the contemporary army-level commander will simultaneously be initiating and waging a main battle. Even as his primary combined-arms force moves by rail to the front, his long-range combat aviation and his advance echelon of motor-mechanized units and cavalry will already be waging fierce battles. In this situation, the reduction of operational activity to the formation and dispatch of columns would amount to bankruptcy. The modern army-level commander must continuously and actively control the course of events, with the step-by-step direction of actions from the depths. Each refusal of active participation in the control process means a step in the direction of operational chaos. The center of gravity within the art of leadership now shifts to controlling the course of the operation and the main battle as was the case in bygone times.

It must be taken into account that there is a time interval between the onset of battle by the mobile advance guard echelon and the introduction into main battle of subsequent combined-arms echelons. There is another interval or pause between the approach of subsequent main force echelons and their entry into the main battle. On a tactical scale this phenomenon was characteristic of Napoleon's era. Now, however, its new quantitative scale determines its absolutely new qualitative content. This pause does not presuppose concentration and disposition of all available forces before battle, as was the case during Napoleon's era. The pause is no longer static. Rather, it is intensely dynamic because of the simultaneous rapid advance of follow-on echelons from the depths even as the advance guard echelon is engaged in fierce battle. However, this pause becomes a reality during the unfolding of a deep operation. Thus, this phenomenon has also closed its circle of dialectical development.

In operational perspective, this pause means that the contemporary army-level commander, with his aviation and his advance guard echelon of motor-mechanized units and cavalry well forward, has the opportunity within his sector to group follow-on echelons and direct them to the front of unfolding main battle. This process will result from the commander's own assessment of the situation and corresponding decision. In the future it will be possible to wage battle in locales decided by the army-level commander, rather than in locales arising accidentally along column march routes, as was the case during the epoch of linear strategy.

Once again the operation becomes controlled.

However, control requires a high order of skill and direction. Diverse data on the immediate situation reflecting operational concerns in two dimensions (along the frontline and in depth) will require a high level of operational art and operational culture to produce an analysis of information and a synthesis of all the elements necessary for a well-grounded decision. In this regard, distant strategic objectives are insufficient. The most immediate tasks must be concretely and progressively resolved in full perspective in order to destroy the enemy throughout his entire operational depths.

Moreover, the intent inherent in decision alone does not form the basis for the art of controlling and conducting an operation. As early as in the late nineteenth century, Lewal wrote that "inspiration was gradually descending from the heights of intellectualism to practical reality," and that "intent was becoming more dependent on available material means." In the future, the main focus of a modern army-level commander and his staff will consist of the choice in methods and organizational techniques for executing an operation. A series of requirements demand great skill in calculation, organization, and direction. These requirements flow from immense scope, huge columns, enormous technological means, and a vast rear.

Control of a modern deep operation means organization first of all.

Contemporary operational art as the art of direction is above all the art of organization.

That is, it is the art and skill of correct calculations, proper organization, and firm direction. The weight of one's own forces and means will lead to great chaos and destruction if organizational art is insufficient or inattentive to organizational detail. It is evident that such art does not proceed from isolated acts of organizational creativity. Rather, organizational art has as its object a field of competence in which everything is closely related, coordinated, and linked. Everything must be decided on some kind of definite and established organizational basis. Modern operational art now approaches a kind of soundly-based concrete system that Moltke failed to foresee. In his instructions to higher commanders, he wrote, "The control of enormous military masses does not lend itself to peacetime study."

Contemporary operational art confronts the urgent necessity for regulating methods of organizing and conducting deep operations with exactness and within the limits prescribed by regulations. In developing prospects for the deep breakthrough, we cannot agree with Lewal's assertions that "imagination and creative work are no longer in vogue," that "flights of fancy are diminishing," and that "the time has now passed for grand preparations, brilliant combinations, and splendid maneuvers." As soon as a breakthrough echelon is released from the depths to conduct maneuver in the depths of enemy resistance, operational art will again confront the challenge of making bold and acute decisions to meet the requirements of the situation.

The army-level commander-organizer who would rationally calculate the control of an operation is that commander who rapidly and acutely perceives the complexity of a situation in all its dimensions, and who immediately makes bold decisions that are both inspired and materially grounded. At this moment, he will be an active commander at his post, since launching combat aviation and hurling the breakthrough echelon into a breach in the front will require direct command. Hours and minutes will be everything. Naturally, such decisions will not emanate from the cozy quarters of an army-level staff located deep in the rear. These decisions will come from an operational command post at the leading edge of the breakthrough. At this post, the modern commander must be close enough to the situation to feel its pulse.

Thus, a modern commander will appear once again on the "Pratzen Heights" [overlooking the battlefield at Austerlitz], but he will be equipped with modern radio and television communications, and he will have an aircraft at his disposal. He will control the deep breakthrough operation with simple hand gestures. A powerful staff, the organizer and technical executor of the commander's decisions, will be at his disposal. The primary operations section will be located at the commander's side. Other subordinate elements of the staff will be located to the immediate rear, where they can control and regulate the advance of units within the deep operational formation. Finally, the logistics element of the staff will be located farther to the rear, approximately along the line of railroad-based deployments, to control the entire complicated mechanism for supplying a deep operation.

So, the system for controlling an operation is itself based on an in-depth structural scheme that inevitably complicates coordination and organization. This control system must be governed by exact regulations. Even so, great art and skill will be necessary to make the system function predictably and accurately.

Finally, when the breakthrough echelon penetrates the depths of resistance to begin its destructive work there, control and decision will reach their highest attainment in the art of encirclement and destruction under the new conditions of deep strategy. After tiresome methods for waging main battles of attrition, this attainment will clearly amount to a renaissance of "grand preparations, brilliant combinations, and splendid maneuvers."

These are the prospects for the evolution of operational art during the epoch of deep strategy.

8. From Theory to Application

We have laid out the prospects inherent in the new epoch of deep strategy in terms of a very general theoretical outline. These terms must now be made concrete in order to translate them from the realm of philosophy and theory into practice. We have tried to define only the basic contours within the new epoch of evolving military art. We have proceeded from the proposition that our future revolutionary-class war will become the greatest act within an armed conflict of world-wide significance. Clausewitz once said that "Each great war represents a separate epoch in the history of the art of war." The emerging epoch of social revolution and revolutionary-class wars will constitute such a new epoch in military art.

But the new forms of military art, which ripen in the process of historical evolution, do not appear spontaneously in concrete manifestations. They must be perceived and studied. They must be philosophically and theoretically substantiated. According to Clausewitz, "each epoch must have its own theory of war, no matter when its philosophical foundations are worked out."

There can be no rational practice without substantiated theory. Therefore, we began with theory so that subsequently we could turn to an examination of the concrete calculus for the deep operation. This approach revealed to us the entire evolution of the art of war since the beginning of the nineteenth century. Only differences between great historical epochs make it possible to discern the internal logic within the development of military art, to explain how and why this art shifted from some forms to others, and to understand why military art finds its culmination in the epoch of deep strategy.

Thus, the single point of Napoleon's era was multiplied during Moltke's era into a series of separate points distributed in space. During the World War, these points were merged into a continuous line. Now this line extends into the depths, producing a square with new spatial dimensions.

We are now entering a new epoch of deep strategy, and we are now making the transition from a broad linear front to a deep front. Of course, the forms of the new strategy will not fully manifest themselves immediately and everywhere. Historical evolution in general knows no firm boundaries. During the initial period of war, prerequisites can still exist within our theater of military actions for a linear strategy of enveloping maneuver. Nonetheless, all the principally new factors of deep strategy will apply to our conflict with full force. This is both because our working class acts as a world-wide historically progressive force, and because the fundamentals of deep strategy are based on the idea of annihilation. The revolutionary proletariat will be the first to employ the new operational art and will produce the first masters of the deep operation for destruction.

As Clausewitz has said, "Actual changes in military art are consequences of changing politics." Only the great political goal of our struggle can assure the historical realization of our deep strategy.

Epoch	Distribution of Operations Along a Front	Distribution of Operations in Depth	Offensive Operational Formation	Forms of Operational Maneuver	Nature of Strategy
Napoleon at turn of 19th century	Single point	Single moment	Deep close-order columns in a single mass	Massed shock-strike along interior lines	Strategy of the single point
Moltke during second half of 19th century	Discontinuous front with points distributed in space	Discontinuous chain of separate, unlinked main battles	Broadly-based distribution in separate groupings	Concentric maneuver along exterior lines	Linear strategy with a discontinuous front
The World War of 1914–1918	Single line extended to form a continuous front	Coherent chain with a discontinuous series of main battles	Broadly-based distribution along a continuous front	Enveloping maneuver along exterior lines	Linear strategy with a continuous front
Future war	Same	Continuous chain of inter-related main battles	Deep distribution in several echelons	Frontal blow along exterior lines and offensive depth-to-defensive depth maneuver for destruction along interior lines	Deep strategy

Figure 5. Evolution of the Character of the Operation.

Source: Original to Author.

Part Three
The Historical Roots of Contemporary Forms of Battle

The basis of battle is overcoming and destroying the enemy's combat formation. Whether the character of an operation is maneuver or frontal, overcoming and destroying the opposing front decides everything on the battlefield. The entire history of the World War proves that weak combat formations in the attack were the basic reason for the appearance of trench warfare and the failure of offensive operations.

In general, tactical factors determine the outcome of an operation. A front can be continuous, without space for maneuver and envelopment, but if offensive tactics manage to overcome resistance, the front will begin breaking up, leading to a war of movement. Or conversely, a front can be interrupted, making envelopment and broadly-based maneuver possible. However, if offensive combat formations fail to break the enemy's resistance, the front will soon become a continuous static wall, as was the case in 1914-1918. Power and assault potential decide everything in battle.

The contemporary attack holds potential for resolving problems inherent in the nature of operations in future war. All the same important factors which show the way for resolution of problems confronting the modern operation lie in the field of tactics. Above all, tactics refract and experience the colossal changes of our time because they constitute both the organizational realm for direct physical action against the enemy and the realm for the direct application of the soldier and his weapon in battle. All the new factors within armed combat bring essential and fundamental change to the field of tactics. This change has manifested itself in our transition to new forms of deep tactics for offensive battle.

According to the essence of these tactics, we have abandoned the gradual and exhausting ways for overcoming firepower-based resistance by stages and by units in favor of the simultaneous penetration and suppression of the enemy's entire tactical depth. With single and simultaneous all-powerful suppressive action, we break up, penetrate, and neutralize the resistance. This is the way to solve the question of overcoming the entire depth of the firepower-intensive front. But, as a new phenomenon, deep tactics for offensive battle did not suddenly materialize out of nothing. Their appearance was subject to the same internal logic governing the development of military art. This process obliges us first of all to understand the reasons and historical roots which elicited and conditioned the appearance of deep tactics. In this regard, the basic point of departure is a critical analysis of the evolution of tactics during the World War. Only history can explain why a given phenomenon assumed a certain character and not another.

1. Tactics before the World War

Two basic and indivisible factors lie at the foundation of tactics as the organization of direct armed impact on the enemy in battle. These factors are the man and his weapon. As an expression of indivisible unity, they determine the power and intensity of the physical force brought to bear upon the enemy in battle.

According to the nature of this physical force, we can broadly divide the history of tactics into two main epochs. The first epoch featured physical impact on the enemy by means of a direct blow, and the essence of tactics was destruction by shock action. The second epoch featured direct combat impact on the enemy by means of firepower, and the essence of tactics became destruction by fire. Napoleon's era might be considered a boundary between these two epochs. Shock force basically governed on the battlefield, although artillery was gaining in significance.

Generally, the epoch of destruction by fire began during the second half of the nineteenth century with the introduction of rifled weapons, particularly the Dreyse rifle and the Krupp rifled field gun. Firepower very quickly became the key material factor of impact during battle. As Moltke put it, the main battle of Sedan in 1870 was fought and won almost solely by artillery fire. Firepower has improved rapidly since then, thanks to growing ranges, rates of fire, and coefficients of lethality.

During Napoleon's era a battalion might yield 2,000 rounds per minute, while during Moltke's era the same figure was 7,000. On the eve of the World War, this figure grew to 11,000-15,000 rounds per minute, and at present it has grown to 20,000.

While the bullet dominated the battlefield, the significance of artillery was growing, although its proportional impact was initially less. Before the World War, bullets accounted for 75-90 percent of targets hit, with the rest left to artillery. Meanwhile, artillery on the eve of the World War expanded to number 6 guns per battalion, thereby becoming an important factor in aggregate firepower along a front line.

This new factor of increased firepower had colossal significance, for it determined the further development of tactics. Above all, it forced a complete reassessment of the power of the defense. During the epoch of destruction by shock action, the correlation of force between the defense and offense was of little importance, since shock-style weapons were equally powerful on both the defense and offense. And Clausewitz, who first examined offensive-defensive correlations as a problem, solved it in the realm of politics and strategy, rather than in the realm of tactics. This problem acquired colossal acuity and importance in the epoch of destruction by fire. As soon as the new rifled weaponry went into action on the battlefield, it was evident that this weaponry was much more powerful and efficient in the defense than in the offense. This realization flowed neither from the qualities or the character of the weapon itself, but from the difference between targets in the defense and the offense.

Indeed, a fire unit sheltered by terrain with attacking targets to the front can be much more efficiently employed than the same unit in the offense facing a firepower system sheltered by the terrain. The superiority of defensive over offensive firepower should have highlighted the problem of how to overcome the new force of firepower by reorganizing troop structures, along with their armaments and tactical forms of action. However, this central question for offensive tactics found no practical resolution before the onset of the World War. On the contrary, Foch wrote before the war that, "The perfection of weaponry leads to an increase in force for the offensive." This was, of course, a wrong conclusion, refuted by the first events of the war.

Before the World War, all regulations and doctrines generally treated the offensive as the only possible way to settle armed conflict. French regulations as edited by Foch were particularly explicit on this point. They stated that "defense leads to failure." This doctrine reached such heights of hysteria that anyone who expressed interest in the defense ran the risk of ruining his career. German doctrine treated the offensive in the same way, but its advantages were expressed in terms of envelopment and annihilation. Still, German regulations devoted significant attention to the force of firepower.

Meanwhile, all this offensive doctrine lacked material and organizational foundations. The composition and armaments of troop structures that went to war in 1914 were too weak to overcome the new firepower. Troop structures were basically a mass of infantry units of uniform organization and armament. Their strength was measured by the number of bayonets. Their armament was a model 1898 rifle that fired a bullet of unsatisfactory composition. In the attack, these infantry troops were absolutely defenseless in the face of fire. They entered the war without even a full supply of properly camouflaged uniforms. An infantry regiment was armed with 6-8 machine guns, which were reckoned more as artillery than infantry armament. Moreover, a cumbersome mount and large shield made these machine guns difficult to handle during infantry combat. Communications required other technological means, and they included 6 sets of telephones and 18 kilometers of wire per regiment. These assets rounded out the complete technological arsenal of the infantry.

The essence of tactical evolution during the second half of the nineteenth century amounted to the lateral dispersion of traditional shock columns into open-order chains and lines that might permit the full employment of frontal firepower. Only at the beginning of the twentieth century did infantry fully migrate to the laterally dispersed open order dictated by the material factor of firepower. However, it is necessary to point out that the open order was not fully acknowledged. The German infantry regulations read, "Frequent reference will occur to dispersal into smaller units and employment of the open order. Rejection of the closed order is wrong and should be avoided." Thus, on the eve of 1914 the continuous closed order was far from a thing of the tactical past.

Transition to the laterally dispersed open order was a natural phenomenon during the evolution of tactics. But the small rifle unit in itself did not have a single element which might make possible the exploitation of its own fire in combat. While providing fire suppression, the larger dispersed rifle-based order lacked sufficient mobility and flexibility both to exploit its own fire and to transition directly to the blow. This fact unavoidably led to protracted fire battles. It also precluded the final stage of combat, tactical pursuit on the battlefield, which the World War never really witnessed. More importantly, the open order as a vehicle for introducing firepower into the offensive was unable to overcome the resistance of opposing defensive firepower systems.

Thus, two very important factors should be noted within pre-1914 infantry organization, armament, and tactics. These factors predetermined the outcome of battles during the first period of the World War. First, infantry composition and armament were powerless to confront and overcome the new firepower. Second, even if infantry managed to suppress enemy firepower, there was nothing in the offensive combat formation to exploit the results and take the battle to decisive conclusion. The outcome of the first battles during 1914 was foreordained.

Of course, artillery should have attained greater significance under these conditions. It was the only force suitable for the suppression of fires resisting the offensive. In fact, artillery had grown in strength and had improved its technical characteristics with reference to range and rates of fire. Still, the proportional weight of artillery in 1914 was quite small. Its shells consisted mainly of shrapnel, with relatively minor destructive force. Of all the various artillery assets in 1914, only the model 1898 German howitzer could be considered a weapon of suppression and destruction in the field. Modified in 1909, this howitzer fired a 16-kilogram shell to a range of 6.4 kilometers. Still, in general, artillery on the eve of 1914 lacked destructive force. The Germans were the first to notice this fact and the first to introduce heavy artillery into their armaments. Corps were allocated 16 heavy howitzers and cannon of 150 and 105 mm.

Tactical views on the importance of artillery corresponded with its technological backwardness. For instance, the French field regulation read: "Artillery fire has only insignificant impact; artillery is an auxiliary arm of secondary importance." The French artillery regulation noted that, "Artillery does not prepare the attack, but only supports it." Only German regulations said that an infantry attack must be prepared with artillery fire. But this commentary assumed the form more of advice and desire than prescription for offensive preparation. In accordance with these views, the normal division- or corps-size artillery complement was viewed as adequate to fulfill all required fire missions. A light 76 mm battery was considered able to fulfill any task along a 200-meter front. So, the established norm became 5 batteries per kilometer of front. Since regulations foresaw a corps in the offensive occupying a 6-kilometer front, the requirement was 30 batteries (5x6), or 120 weapons. Such was the allocation for a French corps, while a German corps possessed even more weapons (160). There was no question either of reinforcement or qualitative improvement. In addition, since a battery could fully perform all required tasks along 200 meters of front, there was no question of centralizing control and massing artillery effects even within the limited confines of the battlefield. The question of controlling and concentrating artillery fire, along with the maneuver of firepower fans, never entered consideration. Even the advanced German regulation never mentioned the establishment of artillery telephone communications. Within the limits of action for an artillery battery, visual communications were deemed sufficient. In this situation, it is evident why advancing infantry, unable to overcome firepower in the attack, failed to obtain necessary artillery support, with the result that the offensive became impossible.

But on the eve of the World War there were other and diverse unofficial views. Schlichting headed a group of military scholars who rejected the possibility of frontal attacks. Schlieffen had based all his teaching on transferal of decision by battle and operation to the flanks. Shallow and deep envelopment were to be the main forms for offensive action. In fact, these views amounted to solving the problem along the path of least resistance. Indeed, the situation in the end of the nineteenth century allowed a certain freedom to pursue and find flanks. But the first events of war in 1914 demonstrated that such views were not sufficiently far-sighted to overcome the evolving nature of operations after the turn of the twentieth century. When fronts became continuous, and when there were no flanks, the frontal attack became inevitable.

Among others, the views of Langlois, a French artilleryman, deserve attention. He clearly advanced the idea that mass artillery fire was obligatory for the offensive. At the end of the nineteenth century, Langlois offered a scheme for organizing an offensive combat formation that was approved only during the second year of the World War. For each kilometer of frontage, he proposed 50-100 guns, without which he considered an offensive impossible.

Unofficial doctrines touched upon another very important question— the powerlessness of the offensive combat formation to exploit its own fire in order to finish battle with a direct blow and tactical pursuit. Bernhardi stressed this question in his *Modern War,* writing, "Under modern firepower conditions, deep deployment is a compulsory element of decisive tactics." Moreover, Bernhardi offered a scheme for such deep deployment. These were, however, only the unofficial views of a military scholar. Still, after the Anglo-Boer War, military literature pointed to the impossibility of deciding battle by way of an attack.

But one should note that such conclusions failed to account for one very important factor: Fire had acquired a new quality and power because earthworks and entrenchments had come to its assistance. The Russo-Japanese War of 1904-05, which the Germans studied very carefully, had demonstrated the importance of this factor, but official doctrine drew no proper conclusions The very same Bernhardi wrote, "No one has yet thought about a real trench war requiring frontal blows." In the final analysis, the nature of battle had been insufficiently studied on the eve of the World War. As the main material factor in this battle, the increased strength of firepower failed to merit full consideration. In the offensive, organization, armaments, and tactics were unsuited to resist and overcome the new force of defensive firepower. Manpower remained the main factor within the offensive combat formation. Little attention was paid to artillery, and battle was envisioned as infantry combat.

Such was the approach of the imperialist armies to the World War of 1914-1918. If one takes into account the reactionary imperialist character of this war, it becomes clear how 30 years before Engels might have exactly predicted that it would be a war "in which millions of soldiers would simply smother each other." This assertion was confirmed by the first events of the war of 1914.

2. The Crisis of Offensive Tactics

The World War began with a brief (1.5-2 months) introductory maneuver period that witnessed a burgeoning crisis in offensive tactics. From the very first battles it became clear that offensive firepower was helpless against the firepower systems of organized resistance. As for artillery, it became evident that a battery deployed to cover a 200 meter front was able to perform its mission only against targets in the open or against an attacking combat formation in the open. However, the frontage norm was far off the mark once artillery confronted firepower systems suited to holding a specific area. As for infantry, in many cases the expert fire discipline of attacking Germans managed to neutralize the fire of English and French resistance. Still, no decisive tactical results could be achieved, since the linear attack order had nothing inside the formation that could directly exploit the successes of fire. The dispersed open order was not suited to this task. The year 1914 never witnessed tactical pursuit on the battlefield because there was no one to fulfill the function. Careful studies of the frontier battles on the Western Front fully confirm this assertion.

In the end, the Anglo-French firepower-intensive front of 1914 was never completely broken anywhere. It simply retreated. The great operational scope of the German offensive to the Marne sometimes conceals the true tactical nature of military events. Actually, the French front (its left flank) had to retreat 400 kilometers not because it had been neutralized by offensive firepower and shock action, but because the Germans had won the struggle for the French left flank. To escape envelopment, the French left flank fell back, pulling the entire French front with it. Denied envelopment, the German offensive bogged down on the Marne. During main battle there, which was essentially a frontal confrontation, the German offensive was essentially helpless in the face of French defensive firepower. After the Marne confrontation, the search for an open flank continued, and the rush to the sea began. This was an event of paramount tactical importance, since the rush to the sea was actually an attempt to elude the bullet. But each time the same bullet awaited the belligerents, because both sides were able to extend their flanks northward. The front rapidly spread laterally to the sea, making a continuous 700-kilometer wall. Thus, tactical factors—the new force of firepower and the helplessness of the offensive before it—conditioned the evolution of operations.

When September-October 1914 witnessed the appearance of the continuous entrenched front, offensive tactics confronted new problems. It soon became very clear that firepower, when employed defensively to hold an area, was highly effective against the offense. Therefore, the side that was forced into defending during the whole course of events in 1914 was the side that first understood the significance of firepower's new force and drew all the necessary conclusions. During the course of the World War, defense led in the search for new forms of battle and new forms for infantry organization and armament. At each stage of development, the defense was superior to the offense. To a certain extent, only the last stage of the World War changed this fundamental relationship.

It soon became clear that the essence of resistance lay neither with the strength of the bayonet nor with its numbers, but with the force of firepower. Already at the end of 1914, the French faced the question of modifying infantry armaments and consequently altering infantry organization. If the ability to hold an area was based on firepower, with the machine gun demonstrating clear superiority in this respect, the material foundation for infantry structure was the machine-gun group, not pure manpower.

The year 1915 ushered in a grand reorganization of troop structures which went almost unnoticed against the background of the war's great events. But in fact this reorganization was very important for the development of military art because it represented a revolution in tactics. In 1914, an infantry regiment consisted of 4,000 troops in 16 companies with 6-8 machine guns. This aggregate might yield a firing rate of 45,000 rounds per minute. After reorganization during the war years, the same regiment by 1918 came to number 1,500 troops in 9 companies with 24-32 heavy machine guns. Further technological innovation produced improved light machine guns that became the foundation for infantry armament. All these elements permitted an infantry regiment under the new table of organization to yield a firing rate of 80,000-100,000 rounds per minute. Thus, although the number of regiments within a division was reduced from four to three, the division's killing power actually increased. This reorganization of troop structure has retained significance to the present day. If we would seek a historical parallel to highlight the significance of this reorganization, then we might compare it with the industrial revolution in England, when the loom replaced the weaver. Within troop structures, the machine gun played the role of the loom.

However, it is necessary to point out one vitally important and peculiar phenomenon. As soon as the infantry was equipped with machine-guns, which might create a barrier in the form of continuous fire lines and beaten zones, the ratio of losses inflicted by infantry and artillery fire changed radically. During subsequent years of the World War, artillery accounted for 50-60 percent of losses. What happened was that the bullet stopped the attacking combat formation and drove it to ground. In this situation, artillery might attack with full force to destroy the entrenched combat formation. Thus, artillery inflicted enormous losses because the bullet had immobilized offensive combat formations.

After infantry rearmament and reorganization, defensive tactics transitioned to new forms. In 1915, German defensive dispositions assumed a belt-like form consisting of three lines with a depth of 4-6 kilometers. Thus, the factor of tactical depth initially manifested itself in the defense. By 1916, a second defensive belt appeared 10-12 kilometers behind the first. At the same time, defensive tactics assumed the nature of definitely prescribed actions. Already in 1915, instructions said it was necessary to fight on more than the first line, for the essence of defensive tactics did not presuppose combat simply on separate fortified lines. Rather, the power of defense lay in depth, for attacking forces that penetrated the first line were to be destroyed by counterattacks from the depths. Thus, views on defensive tactics assumed an identifiable form during 1915-1916.

The same new firepower factor as fully manifested by 1915 would lead to a fundamental review of the basic principles for the offensive, as well as the tactics for the conduct of offensive battle. Upon entry into the war in 1914, the prospects for offensive tactics had received little serious consideration. By 1915, offensive tactics had to face the increased force of defensive firepower and depth of resistance. Contradictions grew between offensive potential and defensive power. The offensive now confronted even more difficult challenges than those that were left unresolved during the maneuver period of 1914. The challenges appeared so unexpectedly that the French could think of nothing better than a return to their former regulations for siege warfare. Such was the grave reality of history.

Under these circumstances, the first attempts at offensive action during 1915 were extremely immature and unconscious. Still, this was the right way to proceed. It was evident that offensive combat formations required thorough and dense artillery reinforcement. From the beginning of 1915, narrow infantry frontages received massive artillery assets, and this factor demonstrated striking evolution until war's end. On the offensive in

1914, 37 infantry divisions with 393 batteries occupied an 80-kilometer front. By the end of the war, the same offensive front was occupied by 75 divisions with 1,432 batteries. Moreover, the proportional weight between light and heavy artillery changed radically (in 1914, it was 11:2, while in 1918, it was 9:7). Already by 1915, the norm of 70-75 guns per kilometer of the offensive front had practically become official.

However, the organization of the offensive itself faced irreconcilable contradictions. The essence of these contradictions boiled down to the fact that the depth of deeply-echeloned defenses exceeded the range of artillery fire in support of the offensive. As early as 1915, complete seizure of a defensive belt required attacking infantry to penetrate defenses to a depth of 6-8 kilometers (where artillery and reserves were located). Soon this depth grew to 10-12 kilometers. Meanwhile, artillery possessed the tactical and technological capabilities to deliver aimed fire only to a depth of 2-3 kilometers within the defenses. The result was an inevitable discrepancy between the infantry's mission to seize a defensive belt and the artillery's capability to suppress defensive targets.

This contradiction could not be resolved, but nonetheless, the offensive required a penetration of 6-8 kilometers into the defensive depths. Initially, the decision was to conduct the infantry attack without artillery support, since artillery was to play only a minor role in the breakthrough. A revival of views popular before 1914 doomed the infantry to bloody failure. A minimal 3-5 hours was allocated to artillery preparation. Artillery tasks were limited to suppression of targets on the forward edge of the defense, thereby leaving the depths untouched. After breaking through the forward edge of the battle area, infantry was to continue the offensive by shock action only. Therefore, the infantry was packed into dense formations at the rate of one soldier per square meter. The density of these infantry formations surpassed that of Napoleon's deep shock columns.

In accordance with this development, the offensive combat formation assumed a new depth. Infantry had entered the World War on the basis of offensive tactics associated with the single dense fire-oriented order. Now, it was acknowledged that the number of defensive lines determined the number of offensive waves. Echeloned defense in depth prompted offensive echelonment. A division's offensive deployment assumed a three-echelon array of regiment following regiment. Waves followed each other at 20-30 pace intervals, with those behind filling gaps that appeared in the first attacking line. In 1915, the first breakthrough attempts resembled a mass scrum. The French conducted their first offensive operation at Champagne in this manner, with tragic results. The offensive covered only

4 kilometers, at which point the French combat formations were simply smashed by German counterattacks from the depths. The Germans fully restored their first line of defenses. Meanwhile, the French lost 45 percent of the infantry taking part in the offensive.

Only heavy losses and a sea of blood forced a reconsideration of questions related to the organization and conduct of the tactical breakthrough. The year 1915 will go down in the history of the art of war as a year of total catastrophe for the infantry. Still, the historical significance of 1915 lies in the events that caused a complete tactical reassessment of infantry's role in an offensive against a fortified firepower-intensive front.

Subsequently, the first instructions to French commanders at the beginning of 1916 read, "Infantry by itself constitutes no attacking force against obstacles. Infantry must never be risked in an offensive against fortified points without a preliminary artillery preparation." So, the fruitless events of 1914 and the bloody failures of 1915 were required before it was acknowledged that infantry could not and must not be thrown into the attack without preliminary artillery preparation.

It became clear in retrospect that battles should be fought not by dense human masses but by masses of artillery and material. This conclusion necessitated both a full reassessment of various elements within the offensive and a search for new technological means. The industrial capacities of the imperialist countries provided enough possibility for the latter. This period witnessed the report of the English Colonel [Ernest] Swinton, who wrote of the necessity for constructing the tank, an idea that was hardly new, since the internal combustion engine had been around for some time. However, the first tank models appeared in the French theater of war only in 1916, although they were not immediately introduced into combat. The same year saw aircraft, which thus far had been employed primarily for reconnaissance, become instruments for direct battlefield application by striking at ground targets from the air. Finally, this was the time that chemical weapons made their appearance. They were first employed by the Germans as early as 1915.

However, during 1915, the widespread deployment of material means for combat found expression mostly within the artillery. Attitudes about artillery changed radically. Indeed, the art of artillery attained a higher level with the massing of fires and the actual suppression and destruction of defensive targets. The most essential consequence of this development was that combat acquired a definite material character that was fully manifested during the subsequent years of the war.

Meanwhile, great and simultaneous changes were also characteristic of the evolving defense. Defense neither lagged nor remained at 1915-levels. On the contrary, evolving defensive tactical forms remained far ahead of the offense. In 1916, the second defensive belt became a common phenomenon, so that the defense came to resemble a fortified zone 15-20 kilometers deep. The defense also assumed a different qualitative strength with the appearance of concrete-covered emplacements that imposed new requirements on attacking artillery.

Under these circumstances, artillery assumed the leading role in the offensive. Main battle became a material expression of combat for destruction and annihilation. This development broadly manifested itself for the first time during events at Verdun.

Although the technological resolution of problems associated with the attack seemed to be on track, tactical organization for a seizure of a defensive belt still faced at least one unresolved difficulty. Its essence lay in the old discrepancy between the anticipated depth of infantry action and the genuine range of field artillery. Resolution of this question simply followed the path of least resistance. Because the maximum range of artillery was 3-5 kilometers, it was incapable of immediately suppressing the entire defensive depth. And, because an infantry offensive without artillery support was now acknowledged as impossible, the attacking depth for infantry was now limited to the range of supporting artillery. Thus, the limits of a breakthrough came to be defined as a 3-5 kilometer penetration of the defensive line. Subsequently, it was considered necessary to reorganize the attack by regrouping and displacing the artillery forward.

As a result, one could count on overcoming the first defensive line only. By 1916, once defenses were 5-20 kilometers deep, it became completely clear that only a series of successive and methodically calculated actions could penetrate a defensive system belt-by-belt. From 1916, the offensive assumed the character of a crawling and exhausting seizure of each defensive belt separately and in progression. In fact, the tempo of advance was 100-1,000 meters per day of combat.

In comparison with the tactics of 1915, this situation reflected a certain degree of progress mixed with regress. Essentially, this method could not be counted on to effect the breakthrough of an entire defensive belt. The events of 1916 were clear indication of this fact. While the defense was gradually and methodically torn apart piece-by-piece, deep defensive tactical reserves remained intact, with every opportunity from the depths to restore each piece to its place in short order. Thus, defensive depth could be maintained permanently, for the entire defensive belt was only falling back, with its depth preserved. It was broken nowhere, and every piece torn off the defensive line was immediately restored from the depth. This realization was of paramount importance for understanding the prospects inherent in a methodical, distributed offensive. These prospects were fully demonstrated by the events of 1916. Today, this realization inspires the idea of a deep and simultaneous neutralization of the entire tactical defensive depth. In 1916, there were no material prerequisites for solving the problem in this manner.

Indeed, conservatism in tactical thinking also played a role. Artillery had already displayed all its might. It demonstrated all the immense suppressive and destructive power which had escaped early-war estimates, despite the fact that heavy artillery had been at the disposal of the German army in particular. Main battle at Verdun, where the Germans attacked a well-fortified position, demonstrated the fact that artillery could neutralize targets. It is interesting to note that on the very first day of that battle (21 February 1916), the Germans penetrated to a depth of 6-7 kilometers, thanks to the destructive force of artillery. This depth was greater than any breakthrough achievement of 1915. The Germans employed the destructive effects of artillery to their fullest extent. In subsequent days during the battle of Verdun, there was a moment (February 24) when the road to Verdun was absolutely open.

But then a new crisis in offensive organization for combat manifested itself. By 1916, it was clear that an offensive combat formation required deep tactical echelonment, without which the kinetic energy of the attack rapidly diminished and then disappeared. In 1916, the Germans formed their assaulting corps into 2 lines, while divisions were deployed in 3 lines. A most unusual density for infantry governed, with divisions deploying on a one kilometer front. Deep tactical echelonment aimed at the quantitative replacement of the leading echelons as soon as their offensive force ran out. In terms of quality, there was nothing different about these echelons, nothing that could promote the exploitation of success achieved by the first echelons, nothing that could press the attack more deeply, while simultaneously transferring the fight to great defensive depth.

The crisis of the events at Verdun lay in the realization that the front line had been breached and that the road to Verdun lay open, but to no avail. The leading German echelons had already absorbed all of the succeeding echelons, and there were no troops left in the combat formation to rush through the breach to develop it in depth. The breakthrough, attained with such great strain of infantry force and with so many artillery assets, could not be developed to a decisive and victorious conclusion. Once this occurred, tactical success on the offensive front in fact became useless, with no real consequences for the attackers, since approaching defensive reserves quickly closed the breach.

In 1916, this very important fact remained undiscovered in the dust of the events. The failure of offensive operations of 1916 produced the general conclusion that decisive offensive operations were absolutely impossible. The first half of 1917 was a period of total tactical confusion, during which nothing new was added to offensive tactics. Tactical thinking remained on the same level. It is interesting to note that this period occurred just as new material means of destruction were making their way into combat formations. In 1917, the French and the English received tanks for the first time, and combat aviation remained actively engaged in the land battle. Meanwhile, complete stagnation reigned in the resolution of offensive tactical problems. The idea was finally legitimized that offensive battle for a breakthrough consisted of a series of successive distributed attacks against precise objectives nominated beforehand. The methodical offensive reached the peak of its development in 1917. True, French directives introduced some very important changes. These directives forbade setting specific objective lines for the infantry assault at some fixed depth. Instead, each tactical echelon should continue its attack into the depths as long as the power remained to do so. This was, of course, a very vague requirement that threatened each echelon with total physical and moral exhaustion. However, in practice, it was found that a battalion might advance 1,000 meters into a breach before the troop complement required replacement. This situation led to further development of offensive echelonment in depth. A four-division corps was generally deployed in two lines, with the regiments of each division attacking side-by-side. Each regiment was deployed in three echelons, with one battalion following another. Thus, an offensive combat formation consisted of 6 echelons of battalions.

The tanks that first appeared during 1917 were to be employed directly with infantry and to be treated only as armor protection for the infantry. The idea that tanks might be employed to penetrate the defensive depth did not even arise. The first battles of 1917 waged by such combat formations produced even poorer results than those of the 1916 breakthroughs. In an offensive waged in the region of Arras, the allies managed to cover only 8 kilometers over 6 days. Moreover, the Germans destroyed 57 of the 132 tanks that were engaged in the offensive. Sixty-four other tanks were damaged, while only 11 managed to return. The tank's debut at the battle of Arras was a total catastrophe.

The second grand allied offensive of 1917 fared little better. The entire operation managed to cover only 5 kilometers. Thus, the first battles involving tanks brought no solutions to the breakthrough problem. Combat was waged only on the leading edge of the infantry confrontational front. Tanks adhered to the same line. The defensive depths remained untouched. Meanwhile, the year 1917 demonstrated more definitively how the defense might employ reserves immediately to restore its defensive belt even when the defense had begun a withdrawal. This was the method that the defense employed to retain its total depth.

3. The Way Out of Crisis

The conclusion of 1917 witnessed great changes in the struggle between the offense and defense, and these changes marked a new stage in solving the breakthrough problem. In November 1917, the battle of Cambrai came unexpectedly like a bolt from the blue. This was an event of paramount importance, even though after three weeks the allies retained only 3-4 kilometers of the breached defensive depth. But this was not the essence of the question. The battle of Cambrai displayed several new ways that the tank influenced the character of the offensive.

These new factors from the battle of Cambrai included the following points. First, fewer infantry forces participated in it; only 17 divisions were deployed in 2 lines across a 12 kilometer front. Initially, divisions received frontages of 2,400-2,500 meters, a marked contrast with the earlier 1,000-meter norm. Second, the battle of Cambrai was initiated without any preliminary artillery preparation. The result was an assault of almost unparalleled surprise, something unknown in earlier offensive practice. And third, offensive combat formations had a cavalry corps deployed within their depths. The assigned mission of the cavalry was immediate development of the breakthrough into the depths of the German defenses. The objective depth for the attacking first echelon and its tanks was set at 10-12 kilometers. On 20 November, the very first day of the breakthrough at Cambrai, the first echelon managed a penetration of 9 kilometers into the German defensive depths. Tanks that tore themselves loose from the infantry appeared unexpectedly in rear areas where German divisional headquarters were located. On 21 November, the road to Cambrai was open, but again there was no one to enter the breach. The cavalry corps rushed from the depth of the attack formation to push through the existing breach, but nothing came of the situation. Passage of the cavalry through that breach had not been properly organized and assured, at which point the course of events in essence froze.

Thus, Cambrai settled only one of the two most essential questions inherent in offensive combat during 1915-1916. In breaking away from the infantry combat formation, tanks penetrated into the depths almost simultaneously with the assault on the forward edge of the German defenses. This fact caused the collapse of the first German defensive belt. But things did not go any farther. The situation could not be exploited because the conduct of battle was not sufficiently organized to assure the passage of cavalry through the existing breach. A tactical breach in the frontline could not be transformed into a breakthrough.

Therefore, the year 1917 did not yield any operational results in the realm of the offensive. But experience revealed some very important factors which convincingly testified to the fact that a defense would fall, provided its depths were simultaneously neutralized. This fact the battle of Cambrai demonstrated, and in this fact lay the battle's enormous historical significance. *The fundamental role in revealing the possibilities for a breakthrough belonged to the tank.* But technological means alone cannot solve a problem unless the art of tactics enables these means to be properly and skillfully applied. But this art was only coming into being. Despite the fact that the Cambrai experience had pointed out a number of factors of paramount tactical importance, the belligerents entered 1918, the year of decisive combat for the imperialist groupings, having drawn no conclusions from the battle. Meanwhile, the allied armies were well-armed with modern materiel. They possessed 2,100 tanks and 6,268 combat aircraft. In contrast, when the struggle was in full swing during 1918, the Germans counted only 1,700 combat aircraft and no tanks. These figures indicated that for the belligerents the great battles of 1918 were waged on absolutely different material and technological bases. Indeed, political factors were different too. Under these circumstances, both major belligerents proceeded from absolutely different material prerequisites in the quest to decide the war offensively.

In 1918, solution to the breakthrough problem was of exceptional interest, for the Germans and the allies achieved it tactically in two different ways. With no other technological means of suppression, the Germans had to base the full material force of their blow on artillery. Their artillery densities reached immense proportions. Earlier, in 1914, during the offensive to the Marne, approximately 5 German batteries were deployed for each kilometer of front. In contrast, at the beginning of 1918, the number of batteries reached 40 per kilometer of front. In 1914, each field gun covered 300 meters of frontage, while in 1918 the same figure was one per 7-20 meters. Thus, total artillery density approached 140 guns per kilometer of front. Mass artillery was the main factor on which the Germans based their breakthroughs in 1918.

Artillery-based methods for suppression and destruction determined offensive tactical forms. Ludendorff's instructions at the beginning of 1918 stated precisely that the entire offensive against a fortified belt consisted of two phases. During the first, infantry was to attack under the direct cover of a moving artillery screen. The screen was to have two dimensions: the explosive fire of light artillery forward, followed by heavy artillery with its large-caliber, high-trajectory fire. It was assumed that the heavy howitzer of the day might reach 5-6 kilometers into the depths, thereby ensuring a breakthrough and destruction of the first defensive belt.

Next came the second phase of the offensive, when combat became decentralized, and infantry on its own pressed forward with the attack, but with a portion of the artillery under its direct control. The echelons within the offensive combat formation were reinforced. During 1915-1916, the German offensive combat formation consisted of 2 lines of divisions, whereas in 1918 it consisted of 3, 4, and even 6 lines of divisions. Ludendorff pointed out the necessity for deep tactical reserves in order to parry defensive counterattacks.

It was peculiar to retain *deep reserves as insurance against counterattack,* when these reserves might better have been used to develop the breakthrough and reinforce a blow into the depths. Ludendorff's emphasis meant that residual defensive tendencies had crossed over into the organization of the offensive combat formation. Once deep offensive reserves were assigned to parry defensive counterattacks, these reserves actually functioned defensively for the offensive, rather than as a means for developing and reinforcing the offensive. This peculiarity influenced German offensive tactics in their entirety, and had an extraordinary effect on possibilities for development of the breakthrough.

In the end, this peculiarity brought about a situation in which the offensive transitioned quickly to defensive tactics, while, in contrast, the defensive transitioned to offensive tactics. Because numerous deep defensive reserves could easily rush to the breakthrough sector, while offensive reserves were being quickly depleted, the main breakthrough battle assumed the character of defenders on the offensive against attackers in the defensive. This phenomenon amounted to one of the great contradictions flowing from offensive linear deployments. However, the tactical formulation of these German deployments attained a high state of refinement.

In accordance with Ludendorff's instructions, divisional frontages expanded to 2 kilometers during the offensives of 1918. Divisions were deployed in 2 echelons with two regiments in the first line and 1 regiment in the second line. This scheme for deployment of the offensive combat formation became the basis for the regulations of most armies. Aviation was widely employed against close-in objectives, and aircraft fully justified their battlefield utility by attacking reserves and artillery positions. However, aviation was not employed as an independent operational element for action in the operational depths. Consequently, the defense enjoyed full freedom of maneuver within its own depths.

The German employment of artillery during the breakthroughs of 1918 is of great instructional interest. German artillery mastery demonstrated the attainment of great skill. A combination of factors, including indirect fire, accuracy of control, flexibility of trajectory, and perfect mathematical accuracy, made an artillery shield possible. This shield constituted a continuous and moving wall of suppressive and destructive fire in front of the advancing infantry. Chemical shells demonstrated a high degree of effectiveness. The practical results of German offensives during 1918 were primarily based on the high organizational attainments of German artillery fire. Suffice it to say that during their breakthroughs in the first half of 1918, the Germans managed to achieve a tempo unknown during the offensives of 1916-1917. The rate of advance for the March breakthrough was 5-12 kilometers per day, while some days during the May breakthrough witnessed advances of 20 kilometers.

In addition to supporting high rates of advance, German artillery managed to achieve full physical destruction of defensive fortifications, installations, and troops. Only 1,200 soldiers of the 157th French Division survived the May offensive. Of the 61st Division, only 800 soldiers survived, while only two companies of the 22nd Division emerged from the May offensive. It is vitally important to remember this situation at the present time, when tanks are being introduced into the structure of the combat formation. Some observers diminish the role of artillery, but it remains more efficient than any other force for the complete annihilation and destruction of a defensive system fortified with obstacles and concrete emplacements. Strong evidence for this assertion comes from the German breakthroughs during the first half of 1918. German tactical attainments are beyond doubt—the Germans achieved tactical breaches of the defensive belt during their first breakthrough operations in 1918.

Designated Operation	Duration in Days	Overall Depth in KMs	Average Daily Advance	Tactical Result	Operational Result
Operation in Picardy	10	Up to 70	7 km	Two complete breakthroughs, 10 & 8 km in width	None, tactical result could not be operationally developed
Operation on the Aisne	10	65	6.5 km	Complete breakthrough, 15 km in width	Same

Figure 6. Results of German Breakthroughs in 1918.
Source: Original to Author.

However, at the very moment when the colossal strain of tactical efforts and mass artillery means had produced a breach in enemy defenses, the Germans had no one to pass through the breach to finish off the attack with operational development of the breakthrough. Operational short-sightedness often leads to situations in which actions become unjustifiable because successful results have not been exploited for attainment of true operational aims. Indeed, it is useless to break down a door if there is no one to go through it. The structure of the German offensive formation did not anticipate the requirement for echelons to develop the breakthrough. Therefore, tactical successes were powerless to bring about any kind of operational decision. Decision retained only a tactical dimension.

The Germans attained offensive tactical success by means of mass artillery fire and maximum exploitation of its technological capabilities. At the same time, this success exhausted the German infantry. In contrast, the allies resolved the problem in quite a different way. The allied way was significantly easier and more efficient because it was based on new technology.

The German General [Hans von] Zwehl coined the well-know phrase, "It was not the genius of Marshal Foch that defeated us, but General Tank." These words were true to the extent that the tank undoubtedly played a vital role in the technical armament of the offensive combat formation. And this situation came about under conditions in which combat application of the tank occurred at an early stage of its development, when it still demonstrated very poor technical characteristics.

In general, the tank resolved two problems. First, it defeated the bullet and thus negated the all-important firepower factor in defensive resistance. Second, the tank combined mobility, firepower, and shock action, and was therefore capable of bringing the full force of combat power into the defensive depths. Actually, solution of those problems determined the solution to the problem of overcoming the firepower-intensive front.

In 1918, the tank was not meant tactically to be a means for destroying the defensive depths. Under conditions of the imperialist war, it was viewed primarily as a means for defending infantry. But the tank's technical characteristics—armor, firepower, and mobility—meant that it was to define its own tactical destiny. Both at Cambrai and in the great breakthroughs of 1918 there was no hint that tanks might be employed as a contemporary long-range group. Even so, tanks managed to penetrate the defensive depths, even if those depths were slight, since penetration depended upon mobility and cross-country capabilities. Tanks broke the defense from the depths and hampered its restoration from the depths. On a tactical scale this factor was of paramount importance to allied success.

Three important circumstances figured in allied success. The *first* essential circumstance was a mass tank attack that sprang from the Anglo-French offensive combat formation. This attack displayed a tendency to penetrate immediately into the defensive depths. The *second* circumstance was infantry deployment across wider fronts, a factor manifesting itself as early as the battle of Cambrai. As a rule, the offensive frontage in 1918 for an allied infantry division was 2.5-3 and even 3.5 kilometers. Broader frontages relieved infantry of the packed and intolerable densities that were necessitated by earlier artillery norms for suppression, but that were unnecessary for reinforcement of infantry shock action. On the contrary, densely-packed infantrymen in tight spaces were unable to utilize their firepower and had to conserve it like dead capital that was vulnerable to destructive defensive fires. Wider frontages in 1918 were also a function of the appearance within allied formations of the tank, which ameliorated the situation, but which did not solve the problem of norms appropriate to an offensive front. However, the evolving tendency to wider frontages was perceptible.

The *third* circumstance was surprise assault without artillery preparation. This circumstance was sometimes carried to harmful extremes. Among reasons for heavy tanks losses (up to two-thirds) during 1918 was the expectation that tanks themselves were to suppress the defensive means they encountered. But in fact no one accomplished this task. In this circumstance, defensive artillery retained great advantage over attacking tanks. Of course, the tank cannot be viewed as an asset capable of resisting any weapon. Such an asset has never existed. The tank's strongest adversary is artillery, and at present it is special antitank artillery. Just as an infantry assault requires preliminary neutralization of machine guns, the tank attack requires preliminary neutralization of antitank systems. This understanding indicates a new and enormous role for artillery during a tank attack, but it was simply not taken into account in 1918. Therefore, tank losses were enormous, but the experience was convincing.

Finally, the *fourth* essential circumstance was the presence within the allied offensive combat formation of a cavalry group following behind the infantry. As an offensive reserve, the mission of this group was to develop the breakthrough to a greater depth, and not to reinforce and support the attack against the front. True, this circumstance did not play any essential role because cavalry could not be properly inserted into the breakthrough. General Debeney wrote, "To achieve greater success in 1918, I tried twice to bring large cavalry formations to the desired place, but despite all their enthusiasm, it took them so much time to overcome all the obstacles that the chance to develop success was lost." Still, the cavalry was a very important factor—perhaps even unconscious—that revealed an attempt to seek an operational solution to the breakthrough problem.

On 18 July 1918, the first allied breakthrough based on the new principles for structuring a combat formation took place at Villers-Cotterets. Approximately 500 tanks took part in the in the offensive to ensure a 10 kilometer advance on the very first day of the attack. By the end of the first day, several sectors in the breached front lay open for passage of the Third Cavalry Corps from its concentration within the depths of the combat formation. The cavalry corps tried to exit the forest in which it was concealed, but managed to deploy only two dismounted squadrons. Clear, easily-achieved tactical success was not crowned with any operational results.

The second great allied tank breakthrough took place on 8 August at Amiens. 680 tanks participated in it. The depths of the offensive combat formation contained a cavalry corps, an armored car detachment, and a bicycle battalion. This group was assigned a mission similar to that of the Third Cavalry Corps at Villers-Cotterets. On the first day of the offensive, attacking forces advanced 12 kilometers, and on the second day, 20 kilometers. A breach opened in the enemy front, but again, the cavalry group failed to penetrate it.

These main battles generally shook the German front, and they resolved the breakthrough problem on a tactical scale. They pointed out real avenues for the development of offensive tactics. But operationally, the breakthrough problem remained unresolved, and it remains a major challenge for our epoch. A poorly-conceived breakthrough echelon played no significant role, because this echelon could not be properly inserted into the breakthrough. Besides, the echelon's composition and mobility failed to meet requirements for breakthrough development. An idea itself is nothing, particularly if it is vague. Only its tactical formulation and its practical implementation can solve a problem. However, in 1918 the allies were far from realizing these conditions. The epoch of the World War brought the evolution of offensive tactics to this stage of development.

4. Fundamental Conclusions

The main reasons for failure in all offensive enterprises during 1914-1918 are becoming clear. In analyzing these reasons one should first bear in mind the class contradictions within the imperialist armies on the Western Front. During the last years of the war, they were on the verge of decay and civil war.

In the realm of operational-tactical forms for the offensive, the reasons for failure lay at first (before the tank appeared) with the fact that neutralization and attack of the defense were conducted only along the front line of direct combat contact. The defensive depths remained untouched. When these depths were finally penetrated by an offensive combat formation, it was weak, demoralized, and bereft of reinforcing echelons. Meanwhile, defensive deep reserves, fresh, freely maneuverable, and strengthened by approaching operational reserves, could each time restore the defense from the depths, assuring its further existence. Under these circumstances, each defensive sector that was destroyed on its forward edge was immediately restored from its depths.

To a great extent, the actions of the attackers resembled the struggle of a knight against a multi-headed dragon. A severed head was immediately replaced by another, and the dragon might be destroyed only if all of its heads were cut off at once. An offensive combat formation was able to destroy at one time only the front heads of the defense. Each severed head was replaced by a new one from the depths, while the defensive body recoiled, remaining on the whole intact and immortal. The situation changed only with the appearance of tanks to penetrate the defensive depths immediately and to deprive the depths of their restorative powers.

The main battles of Villers-Cotterets and Amiens demonstrated that tanks might penetrate into the depths, but that these penetrations remained insignificant. Here was a solution to the tactical breakthrough problem which would predetermine the evolving forms for offensive tactics. During 1918, allied tank breakthroughs demonstrated new ways for destroying the tactical defensive depths. The year 1918 solved the problem of neutralizing defensive firepower by two main means: artillery and tanks. The tank held every advantage for accomplishing this mission. To suppress the defense, artillery had to spend much time and undergo many difficulties, immense technical strain, and colossal expenditures. In contrast, the tank fulfilled the mission while exhausting only ten percent of its overall potential.

Moreover, one should bear in mind that artillery with all its enormous destructive power is capable only of neutralization. It lacks capabilities for the attack. The tank, however, combines firepower, mobility, and shock action, all of which are required to overcome modern defenses. From the point of view of economic efficiency, the tank managed to solve the problem in the cheapest way. According to Fuller, "even at assumed losses of 100 percent, 2,500 tanks would be cheaper than the artillery preparation for the third battle of Ypres taken alone."

Finally, the tank spared infantry all the incredible losses which threatened to bleed it white and make it unfit for further combat. In 1917, occupation of one square mile in the defenses required 8,200 killed and wounded, while after July 1918 allied losses were only 86 soldiers for the same square mile. But one should be very careful about these striking figures. A considerable decline in the German army's combat power, along with its moral decay, figured prominently. However, the facts do testify to an obvious tendency for the growing significance of tanks in determining the fate of infantry on the battlefield. It is quite evident that the tank has become one of the main means of neutralization by resolving at the given moment the task of overcoming firepower resistance.*

*Author's Note: One should bear in mind that there has never existed a weapon proof against all means of defensive destruction. The tank, indeed, has it enemies. First, recent research data testify to the invention of a new ultra-bullet capable of piercing 20 mm armor. The results thus far remain experimental-scientific, and naturally the tank will remain bullet-proof, the victor over the bullet, for a long period of time.

Second, the problem with artillery is different. Results of the battle on 8 August 1918, when 480 tanks out of 650 were destroyed, testify to the fact that prospects for the tank are not favorable in competition with artillery. One cannot underestimate this conclusion. It testifies to the growing significance of artillery, a significance that cannot be diminished. But the essence of the question lies in the fact that artillery faces more difficult challenges at present. During manpower-intensive attacks, artillery's mission in the offensive was to neutralize and destroy machine gun systems. At present, artillery's mission during a tank attack is to neutralize and destroy antitank defensive systems. If this mission is not fulfilled, tanks would sustain heavy losses, transforming their attack into nothing and paralyzing the entire deep battle enterprise.

The tank attacks of 1918 led to no operational result. The reason was that offensive combat formations lacked echelons for breakthrough development. With mobility and cross-country capabilities, these echelons could have developed the blow into the depths to emerge into an open maneuver area behind the defensive system. A tactical breakthrough would then have been transformed into the defeat and destruction of the front. In 1918, the technical characteristics of the tank practically corresponded with this requirement.

However, conservative military-theoretical thought was far from understanding the requirement. Even if the art of organizing the offensive had provided for a proper composition of breakthrough echelons and had ensured their passage through the breach, the final aim of destroying the front would hardly have been achieved. Newly concentrated and maneuverable deep reserves from the operational defensive depths would have confronted units which had just penetrated the breach. The question never arose over how simultaneously to pin down the entire operational depth and isolate the breach in order to preclude the concentration there of deep defensive reserves. Meanwhile, aviation was not assigned missions in accordance with its actual 1918-capabilities.

In sum, the question boiled down to the following: offensive combat during the World War was generally waged along a front line of direct contact, while the defensive depths remained untouched. Tactical echelonment in the offensive served only to reinforce and restore attacking units. Although the tank changed this situation, deep allied echelonment for breakthrough development remained embryonic and played no essential role. Meanwhile, the defense retained full freedom for concentration of fresh reserves within its operational depths. The dominant feature in offensive combat during the Great War was that the combat was waged along a single line of direct contact, yielding what amounted to one-dimensional *linear combat*. Deep echelonment of the combat formation failed to change this phenomenon because deep attack echelons were assigned only to reinforce the combat front line.

Analysis of linear combat in the light of the evolution of its forms during the World War reveals the prerequisites for new principles of offensive combat. To resolve the problems confronting it, offensive combat in light of the war experience must satisfy four basic conditions.

First condition. Sufficient means of neutralization are necessary to overcome the main element within defensive fire, i. e., the bullet. Equipped with firepower and shock power, such means must be sufficiently mobile to penetrate straightaway into the defensive depths. And, such means must be massed to the extent necessary to fulfill the mission through the entire depth of a given sector on the front.

This condition brings us to the problem of incorporating the tank into the contemporary offensive combat formation. In the past, bloody losses were required to demonstrate the impossibility of any attack without preliminary artillery preparation. To avoid unnecessary future losses, we must understand that the contemporary offensive is impossible without the kind of weapon which can resist and overcome the bullet. This weapon must attack the bullet directly with firepower and sufficient destructive weight to constitute a new combined form of shock and attack. The tank is such a weapon.

Second condition. It is necessary to organize offensive combat in order to pin down and neutralize the entire tactical defensive depth simultaneously. If simultaneity is not accomplished, new centers of resistance will materialize from the depths to replace neutralized defensive sectors. The offensive confronts a defensive line that can continuously resurrect itself by drawing upon seemingly endless vital reserves. This situation causes the offense to expend forces and means. If the entire defensive depth is not pinned down simultaneously, the breakthrough becomes impossible. This fact is indisputable and decisive in the tactics of modern offensive combat.

Third condition. The offensive combat formation must have in its depths an operational echelon capable of penetrating the defensive depths immediately following a tactical breach. This echelon must transform tactical success into larger operational results by totally defeating and destroying resistance on an operational scale. If there is no such breakthrough echelon, the simultaneous pinning action and tactical penetration of the entire tactical depth can produce only a pocket-like salient in the breakthrough front. This situation would be more favorable for the defenders than the attackers. All the enormous strain on forces and means would be wasted under these circumstances. This fact is indisputable and decisive in operations for a contemporary breakthrough. Moreover, breakthrough echelons must have different tactical characteristics than those typical of the attack echelons. Breakthrough echelons must be faster. Of course, only motor-mechanized units and mechanized cavalry answer this requirement.

Fourth condition. Even if the first three conditions are fulfilled, they will retain no significance if the breakthrough sector is not completely isolated from the strategic and operational depths of the defense. Without isolation, the enemy's deep combat-ready reserves will rush to the breakthrough sector to greet that advancing breakthrough echelon with a new firepower-intensive front. This response would reduce the development of the breakthrough to naught, and very soon the attackers and defenders would find themselves reversing roles. The vitally important mission for isolation of the breached front belongs to long-range aviation. It must block enemy reserves from access to the breached sector of the front. This aviation must meet and attack approaching enemy reserves at great range to prevent them from reaching frontlines on the ground.

It is evident that ignoring the above four conditions will render impossible any solution to the problem of the contemporary breakthrough on an operational-tactical scale.

The essence of the problem for offensive tactics during the World War lay with linear combat waged along a single line of direct contact. It was necessary to transition to deep combat in order to wage simultaneous battle for simultaneous destruction on several different lines or tiers within the tactical and operational defense. We can then define the character of the new tactics as *the tactics of depth*. These tactics are essentially different in principle from the old tactics of the linear combat formation. But the challenge involves more than a new material basis and new technological means for destruction of the entire defensive depths. These factors comprise only one aspect of the larger problem. In historical and theoretical perspective, resolution of the offensive problem was *absolutely impossible* without transition to the new tactical forms of deep combat. During the present epoch, deep tactics [or deep battle] comprise[s] not only a possible combat form, but also a necessary and inevitable combat form. Without such tactics, no breakthrough can promise success.

Thus, an analysis of evolving offensive tactics during the World War discloses the prerequisites that determined the foundations for the new deep combat. Historically, Fuller was the first to formulate the question of deep combat. In his forecast for a decisive offensive in 1919 (the Entente had not counted on victory during 1918), Fuller proposed simultaneous attacks with tanks against the front edge of defenses and with fast-moving tanks against the tactical defensive depths. The notion of the long-range tank group had not yet been formulated, but Fuller's proposal embodied all the tactical prerequisites.

But his theoretical views went no further. The conditions of bourgeois military theory forced Fuller to reject mass in favor of a theory for small professional armies, which redefines the problem of the offensive in a very different dimension. This theory reflects the class character of an imperialist system obviously in open conflict with the real requirements of future war. For Fuller, deep combat is not a combined-arms phenomenon. His concept of deep combat is absolutely different from ours. For us, deep combat is combined-arms combat, while Fuller's idea "to combine tanks with infantry amounts to harnessing a tractor to a draft horse." Such an idea is unacceptable to us. We will never treat infantry like a draft horse because infantry remains the primary factor within the contemporary offensive combat formation.

In foreign bourgeois literature we cannot find the same concept of deep combat *which already lies at the foundation of our tactics.* Foreign military regulations do not contain any references to deep combat according to our understanding of *the simultaneous neutralization of the entire depths. In this regard, our combat regulations are so far advanced that they constitute the vanguard.* Thanks to an enormous amount of theoretical research and experimentation on new forms for combat, we laid the foundations for contemporary deep tactics.

The forms of deep combat confer new power on the combat formation for overcoming and destroying the firepower-intensive front. These forms create a true and reliable tactical basis for waging the most decisive operations for the most decisive outcomes ever known to military history. On this tactical basis, the wide-ranging maneuver operations of our army are destined to yield victorious results. Operations against an established opposing front will yield an unambiguous breakthrough and complete destruction of the front.

Conclusion

As new forms for the application of modern combat means, deep tactics and deep operations spearhead the assault against the firepower-intensive front. Their chief task is to break and destroy this front through its entire depth. Whatever the circumstance—whether a front is encountered during march-maneuver, during an enveloping maneuver, or during immediate enemy assumption of the defensive—overcoming and destroying any resistance are the basic tasks of deep means for defeating the enemy. These very tasks called them into being. Whatever the nature of resistance, *the penetrating force of the blow* inherent in deep forms of combat is the principal and decisive condition for overcoming and crushing that resistance.

However, it would be wrong to assume that by themselves the characteristics inherent in new forms of combat might either determine the capacity for overcoming any resistance or automatically accomplish enemy defeat under any circumstances. Modern combat means possess great offensive potential and undoubted penetrating power. But in themselves they cannot directly realize this potential and power unless deployed and concentrated in mass in accordance with the principle of attaining neutralizing superiority on selected axes. *No offensive mission can be accomplished without clear and decisive superiority on the axis of the main effort.* The greater the defense's firepower strength and capabilities—an increase in these factors is natural during the historical development of armed conflict—the greater is the requirement for the attainment of power and superiority on the main axis.

Never before has the fate of offensive operations depended so heavily on the main blow and shock groups. Treatment of these two key factors will determine whether offensive operations are able to attain great development in depth or simply get bogged down along the entrenched front. It is singularly evident that any offensive operation begun with insufficient forces and means is destined for stagnation. Each stalled offensive is a step in the direction of a positional front. Therefore, any speculation about distributed fronts, reduced densities, and the possibility of combined blows along various axes contradicts the conditions of modern combat. Such views are wrong in light of an evaluation of the role and significance of modern combat means. *Contemporary technological means for combat increase the efficiency of the blow, i. e., they impart a new quality to the penetration of the blow into the depths. However, these means do not ensure the possibility of blows across a broad front.* Reference to combined blows along various axes is a question of employment densities, and in historical perspective, the whole issue is anachronistic. The conduct of operations with separate groups on various axes was possible during the era of Moltke. At present, a combination of such operations might be inevitable in isolated theaters of war in which dense deployments are impossible. Nevertheless, attainment of a concentrated blow on a specific axis is all the more important in such theaters. However, within primary theaters of war, armies will probably be deployed along a continuous front. A combination of separate operations might be possible during the course of their advance into the depths.

At the outset of war, operations within a theater of military actions will assume the character of an advance of the main force with refused operational flanks and with shock groups on selected axes. The flanks on designated primary axes must have powerful shock groups capable of overcoming any resistance encountered. Otherwise, the offensive will degenerate into a series of disconnected blows incapable of significant results. The commander who wants to ensure the development of offensive operational depth in contemporary war is the commander who concerns himself primarily with the attainment of superior forces and means for the main blow. The composition of these forces and means can never be considered sufficient. For this reason in particular, minor axes should not unnecessarily divert forces and means from the main blow. Otherwise, the entire enterprise will be too weak for the offense and too strong for the defense. Forces and means will be inadequate for the main blow and incapable of deciding anything significant on secondary axes. Any new formation employed on the axis of the main blow will promote victory, while any formation employed on a minor axis will weaken the overall force of the main blow.

Shock groups have a long and instructive history. During the World War of 1914-1918, they underwent radical change. In 1914, the German shock grouping on the Western Front consisted of two-thirds of all forces (35 corps and 10 cavalry divisions) deployed against France. The Germans' enveloping right shock wing consisted of 37 infantry and 6 cavalry divisions that deployed across a 180-kilometer front with one infantry division for each 2.1 kilometers. The stronger the resistance, the stronger and denser the shock groups became. Subsequent years during the World War witnessed a sharp growth in the number of shock groupings and their densities of concentration. In 1914, forces that included 37 infantry divisions, 6 cavalry divisions, 1,572 guns, and 180 aircraft attacked across an 80-kilometer front. In 1918, these same densities increased to 75 infantry divisions, 6,800 guns, and 1,000 aircraft. This is the history of the question. No one can assert that such colossal concentrations will not be required in the future. If a combination of insufficient penetrating force and new defensive means leads to offensive stagnation, then the historical course of events would require even greater concentration of forces and means.

However, such concentrations would be possible only when immobile fronts face each other; the same concentrations would be impossible during the actual course of offensive operations. It is impossible to advance continuously a mass of 75 infantry divisions on an 80-kilometer front. The composition of modern shock groupings must facilitate the continuous forward movement of their combat efforts. This capability has equal significance to the penetrating force of the main blow, and each case necessitates the careful determination of forces and means. In some instances, the two factors of mobility and penetrating force conflict with each other. And, there is unavoidable friction in resolving this cardinal issue. What is evident is the impossibility of consistently moving forward sufficient forces and means required for a breakthrough at a given stage within the development of an operation.

Once again we return to the problem of the deep echelonment of operational efforts to feed the blow from the depths with an influx of new forces and means. The availability of large-scale reserves, especially tanks and artillery, and the deep echelonment of efforts lie at the foundations of deep strategy. The penetrating force is the main factor that determines the potential of a deep operation and its prospects for an offensive advance into the entire depths of a theater of military actions. Overcoming these depths will inevitably require enormous strain.

It is evident within the scope of modern armed conflict that a single operation, even one with the most decisive outcome, will not be able to achieve the aims of the conflict. One or two operations cannot damage the opposition to such an extent that it will be forced to stop the struggle. With all its decisive aims, future war will force belligerents to exhaust all their forces and potential. Conflict cannot be resolved with a single lightning-like thrust. Such an idea would contradict the entire character of future war. As comrade Frunze has said, "As soon as the matter comes down to a serious clash, it will hardly be over within a short period of time after inflicting one crushing blow." Blow after blow will be necessary over a number of stages of intense conflict to attain the final objective of overall enemy defeat. This path to victory will be full of strain and unexplored problems.

Discovery of all the complexities inherent in the depths of titanic conflict is the complicated task of strategic research. This task has never been satisfactorily addressed in the past. At present, the problem appears even more complicated and massive. In his preoccupations with the infinite problems of future war, Culmann wrote, "future war must be waged in whatever way is possible rather than whatever way is desired." But clarity and purpose are incompatible with this approach. The worldwide historical significance of any war imposed on us obliges us to wage it in accordance with the great aims of our policy. Thus, our teaching on strategy confronts the huge task of conducting profound research on the character of future war.

The central question in this research concerns the future character of operations in a theater of military actions. Emphasis will fall on the recurring possibilities for maneuver until attainment of a war's final aim. The deep operation retains in its depths a maneuver of destruction that penetrates into the enemy's depths. But can this operation ensure the development of maneuver-oriented combat during a single overwhelming offensive advance through the entire depths of a theater of military actions? Or, in the final analysis will operations stagnate, as was the case in 1914-1918? Positional forms of combat are scary and repugnant. People recoil from them, as if they were a kind of military plague.

Future war will be and must be a war of maneuver. Such an assertion is either a forecast or a desire which should not be subject to change. However, dogma about the maneuver character of future war cannot solve this great problem by itself; dogma would more surely lead to a positional war. One cannot ignore the World War of 1914-1918, during which the belligerents fought for many years over many kilometers of trench line. Trench war was a grand and cruel manifestation of the forms of the World War. This phenomenon was not accidental. It came about at a particular time as the logical manifestation of a whole series of factors. Today, many observers are inclined to believe that positional warfare has lost its significance. This view testifies to the absence of a proper critical approach to an evaluation of the conditions inherent in modern conflict. A number of factors, including the growing strength of contemporary armies, long fortified borders, and strong defensive capabilities, cannot preclude the possibility of a continuous fortified front. On the contrary, these conditions in many ways foreordain its appearance. At present, the prerequisites still exist for the appearance of opposing continuous fortified fronts; indeed, these prerequisites have even increased in number and significance. Fronts will exist, and they will have to be broken through.

But does this mean that the positional character of future war is foreordained and that maneuver war is impossible? *Not in the slightest.* In positing the question of maneuver war, a huge dialectical error is often made. If all the conditions of armed combat have changed, so also have the contents of its maneuver character. The essence of maneuver actions naturally changes along with the conditions of combat. Not a single phenomenon can be considered static and unchanging in form. At present, the capacity for maneuver is different from what it was in 1914. It is now determined by a different material basis and is now applied against a different operational background. Since there likely will be no vast spaces in future theaters of military action, it is now impossible to base a consideration of maneuver on the availability of open space for freedom of maneuver in future war. This assertion means that there will be no free development of continuous maneuver in its pre-1914 forms. However, under present conditions it would be incorrect to consider notions of positional and maneuver warfare mutually exclusive. The maneuver character of war cannot now be treated in the same manner as free maneuver in free space during a continuous enveloping movement from the Rhine to the Marne or from the Dvina to the Vistula. At present, it would scarcely be possible to conceive of an operation that begins with frontier battles and then advances to cover the enormous distance of 400-600 kilometers. Now, it would be necessary to fight and overcome enemy resistance through the entire depth of the offensive. In these circumstances, the development of maneuver would be possible during an operation within the theater of military actions only if enemy resistance throughout the depths were broken and destroyed each time it appeared.

This resistance might often constitute a continuous fortified front. *However, positional combat depends in essence not on the capabilities of the defensive fortified front, since its appearance is always possible at any given stage of conflict, and even probable at the very outset. Rather, positional combat basically depends on the capabilities and possibilities of an offensive blow. If the offensive blow is always capable of overcoming any resistance that it encounters, then the operation will in general become an uninterrupted advance that assumes a maneuver character in a given theater of military actions. The possibility for the development of maneuver during an operation now depends not so much on the speed of the blow as on its force.* The force of the blow, which lies at the basis of the deep forms of combat, is therefore the fundamental factor.

While shattering resistance along the way, the deep operation essentially presses its efforts into the depths, where it continues to inflict destructive blows. The deep operation is capable of spawning maneuver actions unprecedented in the history of previous wars. Operations during the World War initially unfolded as great enveloping maneuvers that were succeeded by hopelessly insurmountable entrenched fronts. In contrast, the deep operations of contemporary war will be aimed initially at overcoming enemy resistance along a line, and subsequently at the development of broadly-based maneuver actions in the depths. Thus, in comparison with 1914, the course of armed conflict would unfold in reverse order.

The actual course of events, however, will not be simple. Overcoming a single opposing front and then developing maneuver actions into the depths cannot decide the outcome of armed conflict all at once. Efforts pressed into the depths will soon encounter new resistance in the form of a counterstroke or even a new front arrayed in depth. In addition, the initial period of war will surely offer opportunities for broadly-based maneuver actions. This period will constitute a maneuver prologue to the beginning of armed conflict. But this prologue will not last long. Moreover, the maneuver space for fast-moving formations (mechanized, cavalry and motorized) will not be great. Maneuver actions from the line of departure will soon encounter a deployed front, at which time speed must yield to force. Finally, despite limited space, one cannot exclude possibilities for enveloping maneuver; such opportunities cannot be overlooked or refused. Each must be fully, surely, and decisively exploited. Refusal to engage in enveloping maneuver operations at the outset of war—when deployments offer them up—would amount to doctrinaire scorn for those opportunities which historical conditions have preserved.

The essential problem remains that sooner or later—in modern circumstances it will be sooner—an enveloping maneuver will still encounter frontal resistance that must be broken and crushed in order retain the possibility for continued maneuver. One can assume the following sequence of events within the theater of military actions: maneuver actions for invasion during the initial period of war; the overcoming of resistance; the development of maneuver into the depths; and confronting new resistance within the depths. The latter will once again require overcoming resistance, and a new cycle of the aforementioned actions will recur within the depths. These actions will be repeated until attainment of the final aim. Of course, formations will in some measure enjoy various opportunities for maneuver at each stage of the operation. However, the course of operational-scale events within the theater of military actions will generally conform to the internal logic that governs the unfolding of these events within the depths. *And, it is in this unfolding of the operation in depth that one must perceive the maneuver character of contemporary war. Contemporary maneuverability is relationally dependent upon an existing front.* This assertion captures the essential difference between contemporary and past maneuverability. *Contemporary maneuver does not occur in advance of the defensive front (as was the case in time and space during the World War), but behind the front and within its depths.*

Thus, the veil is lifted to reveal the potential developmental contours for operations in future war. It is evident that these operations will impart a new maneuver character to the aim-oriented offensive solely on the basis of the deep blow's penetrating force. But one should bear in mind that combat forms will not demonstrate linear development during a grand and protracted war. As a cruel clash between two forces, war inevitably gives birth to new conditions and new forms. The imperialist armies entered the World War in accordance with the tactics of linear combat and finished the conflict in accordance with the tactics of group combat. Still, the imperialists failed to find ways for waging war that the conditions of 1918 required. Contemporary battle and the contemporary operation will likely be based on deep forms of combat in one or another incarnation.

But one must be sufficiently far-sighted during a war to perceive all its new conditions, and when circumstances warrant, to boldly assume a course for the further development and even alteration of the appropriate forms of warfare. One thing is clear: *At present, the forms of deep combat and the deep operation accord with the logic of history; therefore, they are necessary and inevitable.* Analysis of most of the unsuccessful operations from past wars demonstrates that the gravest errors of commanders proceed not from an ignorance of military affairs. Rather, the gravest errors proceed from lack of familiarity with the general historical development of military art. The result has been application of outmoded forms and methods for armed combat. Linear operational forms are now obsolete, and any attempt to revive them under changed historical circumstances will be a grave mistake.

CPSIA information can be obtained
at www.ICGtesting.com
Printed in the USA
LVHW111515110122
708310LV00008B/756